AF480663

DRIVEN

NO HANDS ACROSS AMERICA, THE
AUTOMATED HIGHWAY SYSTEM, AND THE
BIRTH OF MODERN SELF-DRIVING CARS

BOOK TWO OF DARE TO BE GREAT: A LIFE IN
FIVE ACTS

TODD JOCHEM, PH.D.

with

BARB JOCHEM. M.S.

Foreword by

TAKEO KANADE, PH.D.

Driven: No Hands Across America, the Automated Highway System, and the Birth of Modern Self-Driving Cars

© 2026 Todd Jochem and Barb Jochem

Published by Jochem Family Press
Gibsonia, Pennsylvania and Bloomington, Indiana
www.jochemfamilypress.com

Some names and identifying details have been changed to protect the privacy of individuals.

Any companies or products mentioned or shown here are simply part of the story as it happened; their appearance should not be mistaken for endorsement.

First Edition: 2026

Paperback ISBN: 979-8-9940609-2-6

Hardcover ISBN: 979-8-9940609-3-3

Front Cover: Navlab 5 somewhere in western Colorado or Utah during No Hands Across America in the summer of 1995.

Author Photo: Todd and Barb Jochem, around Christmas 1993 at their home in Pittsburgh.

Cover design: Abigail Franks

 Formatted with Vellum

To Barb —

for her love, patience, and unwavering belief in me. None of this would have been possible without you.

And to my Family —

may this story remind you that boldness and persistence together can carry you farther than you ever imagined.

CONTENTS

FOREWORD

This book tells the lively story of a Ph.D. graduate student in the late 1990s at the Robotics Institute of Carnegie Mellon University. Todd Jochem, a young man from a small Midwestern town in Indiana, was accepted into the CMU Robotics Ph.D. Program — one of the most selective doctoral programs in the country. His early dream had been to play professional football in the NFL; his new ambition was to become a frontier roboticist. Talented, determined, and full of energy, Todd ultimately helped bring to life the 1995 No Hands Across America journey, in which a computer-vision-driven vehicle crossed the country — an epoch-making milestone in the history of autonomous, driverless cars.

In 1984–85, I wrote a proposal to DARPA to launch a research project at Carnegie Mellon University focused on developing vision-based technologies for autonomous driving, as part of the DARPA Autonomous Land Vehicle (ALV) Program. To support this work, we built a computer-controlled van called Navlab (short for Navigation Laboratory), equipped with onboard sensors and computers — Sun workstations at the time — a power generator to supply electricity, and even a long desk with chairs where graduate students wrote and monitored their software.

As in any university research program, graduate students were essential: they developed theories, wrote code, and handled every kind of practical task along the way. One of my favorite jokes in those early Navlab days was that, in addition to sensors and computers, Navlab also carried onboard graduate students — whose safety depended on the very programs they wrote for autonomous driving. It was the fifth-generation vehicle, Navlab 5, that Todd used to carry out the No Hands Across America demonstration with Dean Pomerleau, and I vividly remember joining them for the final leg of the journey, from Las Vegas to San Diego.

Todd writes with energy and humor about his life in Pittsburgh, his academic training, his research toward a Ph.D., and his interactions with a group of smart, interesting, and sometimes quirky people — fellow graduate students, institute staff, and faculty alike. I recognize many of the episodes he describes and recall them fondly. There are others that I did not know at the time, and some where my own recollection differs. What is beyond doubt, however, is that this was a vibrant and formative period in the very early days of autonomous driving research. We did not yet have a clear sense of whether — or when — driverless cars would become part of everyday life as they are today, but we were all deeply excited and optimistic about the possibilities. I am especially proud that I recruited Todd into our Robotics Ph.D. Program and that he became an important contributor to this exciting chapter in the field's history.

This book is excellent reading for young students who are considering careers in academic research — and, more broadly, for anyone thinking about how to build a meaningful, challenging, and exciting life.

Thank you, Todd.

— Takeo Kanade, Ph.D.

Takeo Kanade is one of the most influential figures in modern robotics, but to those who have worked with him, he is equally known for his generosity, vision, and belief in young people with big ideas. As a professor and mentor at Carnegie Mellon University, he shaped the author's graduate experience and encouraged the kind of bold, unconventional thinking that ultimately made projects like No Hands Across America possible. His guidance extended far beyond research; he helped countless students — including the author — see what they could become long before they saw it themselves. Though his technical achievements are legendary, it's his quiet confidence in others that leaves the deepest mark.

SERIES INTRODUCTION

DARE TO BE GREAT:
A LIFE IN FIVE ACTS

This project is my attempt to leave behind something lasting — mainly for my kids and family, but also for anyone who might wonder what drove me, what shaped me, and why I lived the way I did. And maybe provide others the inspiration to break out of what's expected, and live their best possible life!

At its core, it comes back to one truth Henry David Thoreau captured: "Most men live lives of quiet desperation." That line always struck me. We only get one life, and I refused to spend mine muted, stuck in routines, or half-asleep at the wheel. I didn't want to settle for existing; I wanted to live. To me, that meant leaning fully into the gifts I'd been given, taking risks when others hesitated, and pushing myself beyond the comfortable middle ground where quiet desperation hides. I had one life to live, and these principles gave me the tools to live it to its fullest — wringing out as much as possible from my talents, my drive, and the opportunities that came my way.

For me, that has meant holding to a handful of guiding ideas:

1. **Dare to be great.**
2. **Winning matters.**

3. **Success doesn't mean perfection.**
4. **Be accountable.**
5. **Strive for excellence.**

These weren't things I had figured out when I was young. Most of them didn't come into focus until much later — in my 40s, with kids, in the middle of life's demands. And even now, I keep learning — from everyone around me and, now that my kids are adults, especially from them — seeing how they see me, how they see the world, and how that perspective refracts through the lens of my own age. One saying that hits harder with each passing year is this: "I'm closer to the end than the beginning." It's sobering, but also motivating. It reminds me that the time left should be lived fully, with purpose, with nothing held back.

This book series, and the life it reflects, is about striving for excellence, holding yourself accountable, and daring to do things others thought impossible without fear of failure. It's about not waiting for life to happen, but getting in the arena — whether that's on a basketball court in Huntingburg, Indiana, a lab at Carnegie Mellon, or new frontiers of life and technology. It's also about diving headlong into family life, even the uncomfortable parts — the arguments, the stress, the losses, right alongside the joy, laughter, and love. Because a full life isn't just stitched together from the highlight reel; it's made whole by embracing both the good and the bad. That's where fulfillment comes from — not by avoiding the hard parts, but by living through them and letting them shape you.

At its heart, it's a reminder: success doesn't require perfection, daring to be great is worth the risk, and winning — however you define it — comes from preparation, toughness, and never giving up.

If this is the first volume you've opened, welcome.

Each book in Dare to Be Great stands on its own while forming part of a larger story — one life seen from different vantage points: small-town roots, academic breakthroughs, entrepreneurial leaps, family milestones, and the search for purpose. You don't need to read them in order; every beginning carries the story forward. What follows is Book 2 — the season when curiosity met opportunity; when the road from Indiana led to Carnegie Mellon, to a new marriage, to No Hands Across America, and to the first proof that daring to be great could quite literally drive itself across a continent.

SERIES STRUCTURE

- **Book 1: Hoosier Dreams (1968–1990)** – This book begins where it all started, in Huntingburg, Indiana — small-town roots, family, hard work, and the lessons of community life. It's about growing up where the packed gyms of Hoosier Hysteria and Friday night lights shaped both ambition and identity. Just as important was education, where teachers and classrooms opened doors beyond the town limits and made me believe learning could carry me anywhere. Through it all, my parents were always there, steady in their support, giving me both the grounding and the freedom to dream big. In that world, competition, camaraderie, and education together taught discipline, accountability, and the hunger to chase something larger. These were the beginnings — of confidence, of daring to dream, and of the legacy that would ripple forward into parenthood, entrepreneurship, and every chapter that followed.
- **Book 2: Driven (1990 - 1998)** – This book captures the season of life when everything felt new, uncertain, and possible all at once. It was about starting graduate school at Carnegie Mellon, where the pursuit of knowledge collided with the realities of marriage, money, and building a future together. It was about the long nights of research, the challenge of building autonomous vehicles before the world thought they were possible, and the audacity of the 1995 No Hands Across America trip that proved they could be. But it was also about

the beginnings of family life — buying our first house, learning how to build a home, and finding balance in the chaos of academic pressure and personal responsibility. In these years, Barb and I learned what it meant to truly partner in life, to support each other through uncertainty, and to chase a vision bigger than ourselves.

- **Book 3: New Beginnings (1995–2009)** – Life doesn't always move in straight lines — sometimes it leaps forward. These were the years of new beginnings: launching companies from scratch, expanding our family, and learning what it truly meant to be a husband and father. Entrepreneurship brought risk, long nights, and the weight of other people's livelihoods resting on my shoulders. Family life brought joy, exhaustion, and the kind of lessons no degree could ever teach. Both stretched me in ways I hadn't anticipated. This volume is about that stretch — the exhilaration of creating something new, the humility of falling short, and the growth that comes from daring to step into roles I had to learn as I went.

- **Book 4: Full Circle (2009–2021)** – By these years, life had turned a corner. I found myself back on the football field — not as a player, but as a coach and a parent, watching my kids chase their own dreams under the same lights that once lit mine. On Friday nights I wore the headset; on Saturdays I was a cheer dad, standing at the edge of the mat, watching with pride as my daughter's team performed with nerves of steel, discipline, and drive. At home, Barb and I felt the pride and heartbreak of seeing our children grow up and leave for college, even as we began walking with our own parents through the final chapters of their lives. The circle of life was no longer abstract; it was lived in real time — endings and beginnings bound together. This volume is about that circle: the joy of pouring into my kids and their teammates, the gratitude of carrying forward the lessons passed down to me,

and the sobering reminder that time moves faster than we
ever expect.

- **Book 5: Inflection Point (2021+)** – This final book captures
  the stage of life Barb and I are entering — a true turning
  point, where the slope of the curve changes and the years
  ahead take on new meaning. With the kids grown and
  building their own lives, the house feels quieter, but that
  quiet has opened space we haven't had in decades. We're
  rediscovering each other — traveling more, picking up old
  interests, finding new ones, and taking small adventures that
  remind us how much we still love dreaming together. Our
  role as parents has shifted from guiding to cheering as we
  watch proudly from the sideline. And in this new space,
  we've begun asking different questions — not just *what's
  next*, but *what matters most* and how we want to shape the
  years ahead. Inflection Point is about that shift — the
  realization that life doesn't level off here; it bends upward. It's
  clarity meeting momentum, where purpose and gratitude
  begin to accelerate the things that matter most. These years
  aren't a winding down, but a recalibration — a chance to live
  fully, love deeply, and keep daring to be great as the curve
  turns upward and time becomes something to use, not
  manage.

Inflection Point closes the circle, but the story keeps going — in the
lives we've touched, the family we've raised, and the legacy still being
written. My hope is that our family continues adding its chapters —
its voices, its victories, and even its struggles — so that generations
from now, our descendants can look back and see who we were,
where we came from, and what we stood for. In that way, the spirit of
Dare to Be Great will live on — carried forward by those who follow.

AUTHOR'S NOTE

HOW IT HAPPENED, WHY I
DID IT, AND WHO I DID IT FOR

I chose to begin writing my story with this volume, Book 2, because it's the most outward-facing. It tells the story of my time at Carnegie Mellon and my role in the birth of the modern self-driving car. No Hands Across America, the thesis work that followed, and the Automated Highway System demonstration were defining moments — not just for me, but for a field that would eventually change the world. This period of my adult life also left behind the richest written record: papers, news clippings, videos, and stories that can be shared, remembered, and passed down.

But like the rest of the series, this book is written first and foremost for my family. It's my attempt to preserve my story, my values, and the lessons I've learned along the way. I want you to see not just what I did, but why — the values that carried me, the choices that shaped me, and the foundation my early years laid for everything that has come after, including the life we share today. I've always believed that if you see a real possibility clearly — something that could matter — you owe it honest effort, even when the outcome isn't guaranteed.

This book isn't your average memoir. Yes, it's about people, places, and the journey Barb and I took, but it also gets into the nuts and bolts of the technology I created. Not so deep that you need an engineering degree, but deep enough to show how things actually worked, especially in my thesis work. I kept that level of detail for a reason. Part of it is pride — the work described here mattered, and it reflects the depth of what I poured myself into. It had a real and lasting impact, and since none of you were around to see it, I want you to know that outside of our family, it's one of the things I'm most proud of. And while that level of detail may not make the book more commercially appealing, it makes it real.

What I don't always say explicitly — but should — is that none of this would have been possible without a partner willing to absorb uncertainty alongside me.

Some parts of this story may read as if I experienced events with little emotion, just charging ahead. I agree that's often how it looked from the outside. The truth is, I felt every high and low intensely in the moment, but I've never been someone who lingered there. Especially during the most intense academic and high-pressure stretches, I processed stress, doubt, and exhaustion quietly and kept moving forward. I tend not to dwell. I remember very few failures, if any. I put them out of my mind and keep going. That tendency — to keep moving forward rather than relive the past — shaped more of my life than any single success ever did.

Writing this memoir has forced me to slow down and try to name some of those internal experiences. In that sense, it's become a kind of therapy — not because anything was broken (though my kids would probably argue otherwise), but because putting words to emotion is its own kind of growth.

I want to be explicit about something up front: this book would not exist without the help of artificial intelligence.

I have always wanted to record my story — not to write the next great novel, not to win awards, but to leave a clear, honest account of my life for my family and for anyone else who might find value in it. The truth is, without AI, I would never have finished this. Period.

Whether it was one in the afternoon or two in the morning, I could sit down and get my thoughts out of my head and onto the page. The technology was always there — an ever-present ghost editor and sounding board — helping me shape ideas, connect threads, and keep moving when momentum would otherwise have stalled. It brought together perspective and knowledge I wouldn't have had access to on my own, and it met me exactly where I was, whenever I was ready to write.

I supplied the raw material: the memories, the facts, the stories, the digressions, and the long-winded explanations. Technology helped organize, shape, clarify, and refine that material into something readable and coherent.

This is still my story and my voice. The perspectives are mine. The judgments are mine. The emphasis is mine. If you notice an abundance of em dashes, repetitions, or Midwestern understatement — those are on me.

The use of AI didn't replace the work; it made the work possible. It allowed me to do something I had wanted to do for decades but never quite managed to complete. For people like me, that is an extraordinary tool — and one I am genuinely grateful for.

I used AI as a tool to help me tell this story more clearly, but what follows is my truth about this period of my life, as I remember it — shaped by time, emotion, and the imperfect lens of memory.

Finally, the stories in this book only scratch the surface. Supplementary content — including videos, photographs, articles, and additional material that helped shape these pages — is available at JochemFamilyPress.com.

PROLOGUE

A DIFFERENT DREAM

For most of my early life, I believed football was my way forward. In southern Indiana, basketball was king — the true heart of Hoosier Hysteria. Friday nights meant packed gyms — whole towns pouring in to watch, and the game itself woven into the identity of the community. I embraced it fully and was proud to be part of it. But football was the dream I chased. I was a quarterback, and I dreamed the same dream countless kids held: to play Division I football, maybe even find my way to the NFL.

That dream ended on one play against Tecumseh High School on a cool October night — October 11, 1985. We won the game 21–13, but for me, the season was over. On an option to the right, I was hit high and low at the same time, and my fibula snapped. As I was being tackled, I could tell something wasn't right — it just didn't feel normal. I looked down and saw the bone pressing against the skin, my foot twisted at the wrong angle. I remember feeling scared. It didn't hurt at first, but I knew something was wrong and didn't know how to process it. I just rolled onto my back. I don't think I yelled, but I might've told my teammate I needed help. Surgery put a plate in my leg, and thirteen weeks later, even though I was still limping, I was cleared to be back on the basketball court.

That senior basketball season turned out to be one of the best of my life. Our team went on to have another memorable run that I was proud to be part of. Returning from the injury to contribute was both redemptive and unforgettable. But as far as football was concerned, the damage was done. Division I recruiters stopped calling, and although I didn't want to admit it, I knew the NFL was no longer in the cards.

At first, I felt a little lost. This was the first major injury I had ever had. When you have spent your whole childhood imagining yourself under the lights, what do you do when those lights go dark?

For me, the answer — though I didn't realize it then — was to lean into academics. At Indiana State University, during my junior year, I began shifting my mindset. School had always come fairly easily, but now I began to treat academics the way I once treated football: as the arena where I could compete, prove myself, and imagine a future. That was when graduate school first began to take shape in my mind.

Then came the call that changed everything.

It was from Takeo Kanade — one of the world's foremost roboticists and the director of Carnegie Mellon's Robotics Institute. He wanted me to come to Pittsburgh — he was giving me a shot.

I can still remember how improbable it felt. A small-town kid from Huntingburg, Indiana, who'd gone to Indiana State — not MIT, not Stanford — being invited into the very heart of robotics research. I didn't fully understand what it meant, but I knew it was a door swinging wide open.

Talking to Takeo wasn't easy for me. His accent wasn't particularly strong, but to my Midwestern ears — untrained and unworldly — it took effort to follow. I wasn't yet comfortable with the diversity of voices and backgrounds I'd soon encounter at CMU. But I knew enough to say yes. I didn't know what was ahead, but I sensed my life was about to break from the familiar Midwestern script.

Years later, I asked Takeo why he had accepted me into the Ph.D. program. I was from Indiana State — a school he probably hadn't even heard of — while most other applicants came from elite universities. His answer was simple, almost disarming. He said, "I saw that you were a football quarterback, and I asked around if the football quarterback had to be smart. People told me yes — so I figured you could make it here." It wasn't the answer I expected, but in hindsight, it made perfect sense.

When I first arrived in Pittsburgh, I expected a dying steel town. That was its reputation — Rust Belt grit, old mills, smokestacks. But that wasn't the reality. Pittsburgh was already reinventing itself as a tech and medical hub. CMU and the University of Pittsburgh were part of that resurgence.

Still, to me, it felt foreign. I had grown up in a small Midwestern town with ranch houses and wide streets laid out in a square. Pittsburgh was different: steep hills, winding roads, neighborhoods that didn't follow any grid. Barb and I were married now and lived in a row house with no front yard, a cracked foundation, and on-street parking. It was a world away from Huntingburg.

Yet it wasn't the East Coast either. Pittsburgh shared something essential with the Midwest: people valued hard work, loyalty, and community. The locals — Yinzers — were quick to accept us once they saw we were good people willing to work. And the funny thing was, both the Pittsburghers and the other Americans at CMU thought we were from the South, because of our southern Indiana drawl. In a way, we were outsiders twice over.

Now I see how much of this was preparation. I had already learned resilience on the playing field, curiosity in my tinkering, and a bit of toughness from my small-town roots. Coming to CMU tested all of that at once.

The NFL wasn't meant to be. Carnegie Mellon was.

And this book is the story of that evolution — from small-town Hoosier to graduate student in Pittsburgh, from fumbling with compilers to guiding a minivan across the continent with no hands on the wheel. The chapters ahead trace that improbable journey — from broken bones to breakthrough moments, from small-town dreams to the most famous road trip in the history of robotics.

INDIANA ROOTS

SPORTS, CURIOSITY, AND A
SMALL MIDWESTERN TOWN

"I'll make it."

— JIMMY CHITWOOD, HOOSIERS

That simple line captured everything about what it meant to grow up with Hoosier dreams — belief, pressure, and the quiet confidence that if given the chance, you'd rise to meet the moment.

This chapter is a fast-forward recap through my young life. It may feel scattered at times, but it gives you the context — the people, places, and moments that shaped me before Carnegie Mellon.

HOMETOWN AND FAMILY LIFE

Huntingburg, Indiana, was a small town with a big heart. Everyone knew everyone, and life revolved around school, church, and sports. It had been around for over a hundred years by the time I was born, shaped by both its geography and its people. Major roads and a railroad line ran through town, connecting everything but never making it feel any less like a tight-knit place. In true Midwestern fashion, the east–west streets were numbered — streets or avenues — while the north–south ones were named after trees or presidents.

Everything centered on 4th Street, where the mom-and-pop shops were clustered. At either end of town sat the grocery stores — one was an IGA, the other an A&P. When I was growing up, there were only two stoplights, and on one of those corners, the bank had a sign mounted to its building that rotated and displayed the time on one side and the temperature on the other in glowing lightbulbs. We always wondered what it would do when the temperature hit triple digits.

I don't really know what it was that we Midwesterners — or maybe people in general back then — had about needing to know the time and temperature, but we even had a phone number you could call — from our party line telephone connection — that told you both. I think it played an ad for one of the local businesses, too.

Even though it was a rural community, and farming played a central economic role, there was also a significant amount of industry — mostly wood furniture manufacturing — and that helped the town maintain a solid middle-class base.

I think my extended family was close in the way most families were back then. My parents and their siblings talked a lot, and I know there was a strong bond between them. But as one of the youngest cousins on both sides, I was always a little outside the circle. By the time I was old enough to pay attention, most of the others were already grown or off doing their own thing, so I never really formed strong ties with any of them. Still, like most families at the time, we all got together for Christmas and Easter, and all of my aunts and uncles — except maybe one — lived somewhere in the county.

I was an only child, and both of my parents worked — part of a gradual shift in our town from single-income households to dual-income families. My dad worked in a factory his whole life and was soft-spoken and easygoing. My mom worked at the Farm Credit agency and was the more structured one — not strict exactly, but she had clearer rules and higher expectations. I always knew they both loved me deeply and would do anything for me. Like many kids of that era, I wasn't exactly a latchkey kid, but I did have plenty of independence. I was responsible — I didn't run with the wrong crowd (if there even was such a thing in Huntingburg) — but I was also a typical only child: independent, sometimes self-focused, and used to being the center of attention.

I grew up with the kind of Midwestern childhood people look back on with quiet envy — though at the time, it just felt normal. My grandma was my babysitter, and in the summer I'd take off on my bike first thing in the morning and ride all over town. I didn't have to check in — just be back by the time my mom got off work. We organized pickup baseball games in the dusty grass lot behind the old hospital, met at the city park for summer camp, swam until we were sunburned, and lived like time didn't exist. We'd stop by the snack shack for a cup of Coke and whatever candy we could afford, then head right back out into the heat. I didn't have a care in the world. It was the kind of freedom that only feels ordinary when you don't yet know how rare it is.

LEARNING BY MAKING

My parents supported my tinkering interests. One of my earliest favorites was the Science Fair 150-in-1 Electronic Project Kit — a big plastic board full of spring terminals, resistors, capacitors, transistors, lights, and a project manual overflowing with possibilities. I'd sit on the floor wiring up little circuits: a light bulb, a crystal radio, a Morse-code oscillator. Watching a bulb flicker on or a buzzer sound felt like magic — you could take a handful of parts and make something come alive.

Model rockets came after that; I built dozens of them, mostly from Centuri and Estes kits — from foot-long beginner models to multi-stage rockets and towering three-foot builds. I even started a Saturn V once, though that project proved too daunting for my eleven-year-old self. Launches were their own kind of rush: press the button, the fuse hissed, then *whoosh* — the rocket leapt skyward until the orange-and-white Estes parachute deployed. Except for the times when I packed the parachute wrong or jammed the nose cone too tight — then the rocket returned like a ballistic missile. Those thrills echoed the excitement I'd later feel watching robots and cars come alive in the lab.

I also built remote-control cars from Tamiya and Kyosho kits. I thought the 4-wheel-drive models were incredibly cool, but it wasn't the mechanics that fascinated me most — it was the circuitry and radio-control systems. Those projects may have been early seeds of my interest in robotics. The idea that you could send signals and make something respond — not just with gears and wheels but with logic — grabbed me in a way wrenches never did. Those evenings spent hunched over instructions and wiring little motors turned out to be early training for a career in engineering.

CURIOSITY WITHOUT LIMITS

It didn't stay confined to electronics. When the Three Mile Island nuclear power plant accident occurred in 1979, I became fascinated with how reactors worked. I remember vividly sitting on my grandma's couch, eyeballs glued to the TV watching updates about what was happening. I was just a kid, but I pored over every article I could find and drew the system again and again — how heat from the reaction turned water into steam, steam into turbine rotation, and turbine into power. I'd tweak my drawings until they matched what I saw on the screen. I wasn't afraid to dive deep. When something caught my attention, I wanted to understand it inside and out. I was so captivated that I even completed the Boy Scout Atomic Energy merit badge during that time.

DISCOVERING COMPUTERS

The first time I saw a computer, I was hooked. It was a Radio Shack TRS-80, sitting in my sixth-grade classroom at Huntingburg Middle School. The TRS-80, introduced in 1977, was among the first mass-produced personal computers. It had a black-and-white monitor, a clunky keyboard, and used cassette tapes for storage. No frills. But to me, it was alive. The teachers let me use it during study hall and at lunch. I wrote simple games in BASIC — text adventures, number-guessing challenges, and a game called "Kill the Moose," where little "rockets" shot across the screen at a "moose." Of course, the graphics looked nothing like rockets or moose — just ASCII characters, but to me it was magic. For the first time, I realized you could tell a machine what to do and it would listen.

By high school, we even had a computer programming class — unusual for the early 1980s. The school used Educator 64 machines, and for the first time I had real assignments to write programs in BASIC. It wasn't just tinkering anymore; I had to make the code work, hand in results, and see my programs graded. That was the first time programming felt like a discipline, not just a novelty.

Around the same time I started showing interest in programming at school, my parents bought me my first home computer — a Commodore VIC-20. The VIC-20, released in 1980, was the first computer to sell over a million units. It had just 5 KB of RAM — barely enough to hold a page of text. You saved programs to cassette tape, and if you forgot to save, everything disappeared when you turned it off. Primitive, but utterly captivating. After that came a Commodore 64 with more memory and better graphics, opening up even more possibilities. And somewhere in the mix, I also got an Atari 2600 for Christmas. Watching what professional programmers could do with cartridges fascinated me — even if you had to blow the dust out to get them working. I didn't have the words for it then, but something about messing with those early electronic devices pulled me into the zone — where time fell away, my focus locked in, and every little breakthrough made the next one come faster. It felt like pure momentum.

In college at Indiana State, I upgraded to a Macintosh Plus, a graduation gift from my parents. At first I used it mainly for word processing, but I also had an account with America Online (AOL). Back then, AOL wasn't the Internet we know today. You dialed in with a modem, listened to the screech and hiss of the connection, and suddenly you were inside a sprawling online service with chat rooms, message boards, and searchable libraries. To me, it felt like a vast electronic garden of information, a true precursor to what the Internet would soon become.

What I'll always remember, though, was a late-night project in the school's electronics lab where I wired the Mac Plus to a microcontroller and an LED. Whatever I typed on the Mac would flash across the LED. It was crude, but it worked. One detail sticks with me: the original Mac didn't use the common RS-232 serial standard but RS-422, a faster, differential signaling method that Jobs and Wozniak had adopted to push past the baseline. I had learned how to use it with chips and coding, and I felt a real sense of accomplishment.

I'd lug that Mac Plus across campus to the Electronics and Computer Technology building at night after football practice, feeling a little embarrassed but also a little different — and, to some extent, a bit proud of myself for doing something the other guys weren't doing. Once I got there, I'd spend the evening soldering and coding while my teammates were out doing other things. I don't think I really believed any of it was going to move the needle in terms of where I'd end up in life. I was just curious if I could make it work, and it was fun to me. That little demo — a blinking display controlled by a computer — felt every bit as thrilling as launching those model rockets years earlier.

WHERE THE TOWN GATHERED

Still, in Huntingburg, technology was an unusual sidenote. Basketball was king. At the center of that world was Huntingburg Memorial Gym — The Big Gym. Built in the early 1950s by local businessmen to rival Jasper's, it held over 5,000 people, and 6,200 when packed for sectionals — both more than the town's population.

Walking into that gym during sectionals felt electric — bigger than our little town had any right to be. The smell of popcorn and pickle juice hung in the air. Bright lights lit up the court like a TV stage. The roar when we charged out of the locker room shook the rafters. To even get a sectional ticket, you had to buy a season ticket. It was that big a deal.

And it wasn't just the gym. One of the highlights of the year was the Fourth of July YMI picnic at the city park — games, food, and fireworks that brought the entire community together. Those kinds of events made small-town life feel bigger than it was, and gave me a sense of belonging that has always stayed with me.

PLAYING FOR COACH DUNCAN

Head Coach Gary Duncan ran our basketball program, and he was more than a future Hall of Fame coach — he was a mentor and a leader. One of his boldest decisions came early: he pulled me and a few others up to JV as freshmen and cut several upperclassmen. He told them the truth — they could stay on the roster, but they wouldn't play. It was harsh, but it cleared the way for us younger players, who had far more potential. Watching him handle that moment taught me the value of making tough calls for long-term success.

He pushed us relentlessly but also told us he loved us, which was rare for a coach at that time. Our first major success came when we beat county rival Forest Park handily in the Sectional finals, then went on to win our first-ever Regional. In the Semi-State, we upset L&M — a team featured in *Sports Illustrated* — by 18 points before falling in the State Finals to powerhouse Marion.

Our senior year brought another unforgettable run. We beat Jasper in the Sectional finals in a true barn burner — both teams shooting better than 60% — and then captured the Regional and Semi-State to earn another shot at Marion. That Marion group, known as "Purple Reign" for their colors and talent, was a full-fledged dynasty. They beat us again that year, on their way to a second consecutive state championship, and repeated the feat yet again the following season.

I'll never forget sitting in the locker room after that final loss, watching my teammate Ron Patberg cry, and seeing Coach Duncan kneel beside him, offering comfort. Those were the last moments we wore those jerseys, but the lessons stayed. I was upset too — not just because we lost, but because I knew, even if I couldn't quite articulate it then, that this was the last time my friends and I, the guys I'd grown up with, would be truly competitive at something together. Baseball and track were still ahead in the spring, but they didn't carry the same weight. This felt like the end of a chapter, and I could feel it settling in, even as I tried to shake it off.

FOOTBALL AND THE INJURY

While basketball defined Huntingburg, football defined my dreams. I was a quarterback and thought I might play Division I ball, but after I broke my leg, Division I recruiters stopped calling. It was devastating. Once the surgery was done and the cast came off, I didn't have anything left to focus on except recovery, so I threw myself into it with everything I had. Getting back as fast as possible became the only goal that made sense. What I didn't fully grasp then — though I could feel it around the edges — was that things were already shifting. The future I'd imagined for myself in football was starting to tilt in a different direction, even if I wasn't ready to admit it.

That pivot didn't fully come until a few years later, when I was a junior at Indiana State. But looking back, I can see the beginning of it right there in that rehab — the same drive that had pushed me on the field: preparation, toughness, repetition. Those patterns were already forming, long before I knew they would one day carry me through grad school and into an entirely different life.

BARB

I had known Barb my whole life, but I didn't really *notice* her until the spring of my senior year in high school. She was absolutely gorgeous — the classic Midwestern girl with a tomboy streak from keeping up with all her brothers. Blonde, athletic, and carrying a natural, effortless confidence. You can dress it up or just say it plainly: she was drop-dead beautiful.

But it wasn't just her looks. What really caught me was her personality. She was confident around me, pushed back when I needed it, and had a warmth and kindness that showed up instantly. Those traits pulled me in then — and, in different ways, still drive me crazy today.

I asked her out — and you'll have to read Book One to get the full story — and even though she had a few reservations, she said yes. And despite heading off to different schools — me to Indiana State and Barb to Indiana University — we stayed together all four years and got married right after graduating. I was 22 and she was 21.

We were married at her church, which sat at one end of Main Street, while my church was at the other — and it felt like everyone in between came to the ceremony or the reception. The reception was at the Holiday Inn in next-door Jasper, complete with beer, a DJ throwing colored lights across the room, and plenty of '80s music. Friends and family came from everywhere.

It was classic Southern Indiana, through and through: a packed church, fried chicken with corn and green beans, pies and cakes, and everyone celebrating under one roof. After more than 35 years, I'm still amazed at how young we were — and how natural it all felt. Not premature. Just right.

MENTORS BEYOND SPORTS

Sports shaped me, but so did teachers and others who saw something in me. Mr. Menke, my high school physics and calculus teacher, was also a farmer — a man who worked the land but lit up just as much when he was talking about science. He was one of the few people in my world who shared my curiosity about how things worked, and he was genuinely smart. For a small high school of about 500 students, I was lucky to have a teacher like him. He challenged me in ways I didn't think possible, showing me that math and physics weren't just equations but keys to understanding the world — and that aiming higher, even in a small place, was possible.

At Indiana State, Dr. Cockrell — who was also my advisor — played a similar role. He saw potential, encouraged me, hired me for a part-time lab job, and wrote the recommendation letter that helped me land at CMU. At the time, I didn't even know there was such a thing as graduate school. Without mentors like him, I never would have known the path existed, much less had the opportunity to take it. At a big school, I might have been lost in the shuffle. At ISU, people like him noticed and invested in me, and in their own way pushed me to dare to be more than I thought I could be.

I also found my first mentor outside of sports and school — and beyond my dad — at Indiana State, someone who opened my eyes to a wider world. His name was Ed Pease, our fraternity's chapter advisor. He was less than twenty years older than me, close enough to feel relatable but far enough ahead to serve as a model. His path was one I had never seen up close: a fellow Hoosier who went on to Indiana University, built a career as a lawyer, and later served in Congress. What struck me about Ed wasn't just his résumé, but the way he wanted our fraternity to be different. He believed the fraternity could stand apart from the stereotype, and that idea stuck with me. For me, that was the first glimpse that greatness wasn't limited to the field, court, or classroom.

Among the farmer who loved physics, the professor who believed in me, and the congressman who modeled leadership, I had been handed a quiet but powerful roadmap for what was possible. Each of them pulled me a little further out of the small-town bubble and pointed me toward a larger horizon.

REFLECTIONS

My childhood and youth weren't just about sports glory. They unfolded along parallel tracks. On one side, the Big Gym, Coach Duncan, and the pursuit of championships built toughness, resilience, and lessons in leadership. On the other, electronics kits, rockets, RC cars, and the TRS-80 fed a growing hunger to understand how things worked — shaped by mentors like Menke and Cockrell. And Ed Pease was the one who pushed me beyond the boundaries of small-town thinking.

During most of my years at Indiana State, my ambitions were modest and concrete. I wasn't chasing anything abstract or grand. I expected to find a technology- or engineering-related job paying around thirty thousand dollars a year, buy a decent ranch house, and build from there. That was the plan, as fully formed as it ever became. I didn't imagine staying in Huntingburg, but my sense of geography didn't stretch far either — maybe Indianapolis, maybe Louisville. Chicago, St. Louis, or Cincinnati already felt distant. Silicon Valley wasn't in my mind. Neither was research, or robotics, or a PhD. I wasn't restless or dissatisfied. I assumed the future would unfold in a straight line, one reasonable step at a time. Carnegie Mellon hadn't entered the picture.

I was fortunate. My parents encouraged my tinkering. My teachers welcomed my questions. My coaches taught me grit. And just as importantly, I found an incredible partner in Barb — someone I could grow with, who made every version of my future bigger than the one I could have reached alone. All of it came together to shape someone who would later thrive in the chaos of graduate school at Carnegie Mellon.

Hoosier dreams, yes — but not just sports. Dreams of daring to do hard things — on the court, in the lab, and in life. In time, those Hoosier dreams would grow beyond Indiana, carrying me to Pittsburgh. I didn't yet know what shot I'd be asked to take, but like Jimmy Chitwood, when the time came, I knew I'd say, *I'll make it.*

PITTSBURGH AND CMU

FROM HUNTINGBURG TO
THE ROBOTICS INSTITUTE

That spirit of persistence and seeing opportunity in every challenge — even when you don't yet have a map — is something I carried with me into the next big crossroads of my life: choosing where to go to graduate school. Before I could arrive at CMU, I had to choose a direction.

CHOOSING CMU

Choosing CMU wasn't just choosing school — it was a leap into the unknown. And back then, applying to grad school looked very different. There was no Internet, no online portals, no email blasts from schools trying to recruit you. If you wanted to know what programs were out there, you went to the departmental bulletin board. Tacked to the cork board were flyers advertising graduate opportunities, each with a row of pull-off tabs with phone numbers. You tore one off, called, and asked them to mail you a formal paper application. That was the process.

I applied to the Ph.D. programs at Michigan and Florida — with Florida more of a safety school and Michigan the one I hoped to attend. And then there was Carnegie Mellon's Department of Electrical and Computer Engineering, clearly my reach school. (The Robotics Institute hadn't formalized its Ph.D. program yet.) On CMU's application, there was a small box marked "Interests." In it, I wrote exactly four words: "Robotics and Computer Vision."

Those four words changed everything. Takeo Kanade — already a legend, though I didn't know it then — was serving on the ECE admissions committee. He saw those four words and picked up the phone to call me. I wasn't home, so when I got back, I found a handwritten note from my roommate taped to the door: "Takeo Kanade called." The name didn't mean much to me then. Only later would I realize the weight it carried.

When I returned the call, Takeo got straight to the point. Would I consider switching into the new Robotics Ph.D. program instead of ECE? And then the unforgettable part: if I wanted the spot, I had to decide by the next morning. No deliberations, no long deadlines — just a yes or no.

When he said that, I felt butterflies in my stomach — the excitement of being offered the opportunity, but also the gravity of making a life-shifting decision on gut feel alone, with little more than the reputation of each school to guide me. I knew I would disappoint Barb if I chose Carnegie Mellon, but I also knew it was likely the best long-term outcome for both of us. It wasn't an easy decision, but in some ways it felt obvious. Deep down, I knew where I had to go.

The rush was quaint by today's standards. Takeo was leaving for Japan the next day. With no Internet or email and no quick back-and-forth, he needed my answer before he boarded the plane so he could manually push my admission paperwork through. If I didn't call back in time, the opportunity was gone.

It felt like a coin toss, but in reality there was no real question. Michigan was the safe bet, and we already had close college friends living near Ann Arbor. Florida was the comfortable option, with Barb's sister in Gainesville. But Pittsburgh? We had no one there, had never even visited, and knew almost nothing about the city. CMU was the frontier, so I said yes. And those four words — Robotics and Computer Vision — became the origin story of my adult life. They set the trajectory not just for my career, but for the rest of our lives. That decision didn't just affect me — it changed Barb's world too.

Years later, after I had established myself and was doing meaningful work, I finally asked Takeo about it directly. At a lab retreat, I pulled him aside and asked point-blank: "Why did you call me? Why did you take a chance on me when there were so many other applicants from powerhouse schools?" His answer was both shallow and deep. He said, "I saw you were a college quarterback. I asked people if quarterbacks had to be smart. They told me yes — that quarterbacks had to be able to control the whole offense. So I decided to call you." It was classic Takeo: deceptively simple, almost offhand, but with a logic that ran deeper the more you thought about it.

Barb's Perspective — On Choosing CMU over Michigan or Florida

When Todd chose CMU, I'll admit I was disappointed at first. With our families and friends in Michigan and Florida, the thought of moving somewhere we didn't know anyone felt daunting. But pretty quickly I reframed it as an adventure. Exploring Pittsburgh, driving through neighborhoods, and looking at houses became exciting. It was unfamiliar, but it also felt like the first real chapter of our story.

In retrospect, I think moving away from family made us stronger as a couple. When things got tough, we didn't have anywhere to run — we had to turn toward each other and make things work. Plus, we relied on each other for fun, entertainment, and meeting new friends.

That one decision — sealed in a single phone call — carried us from the comfort of the Midwest into the unknown of Pittsburgh, where the Robotics Institute would become my proving ground and where Barb and I would start building our life together from scratch. But first, we had to get to Pittsburgh. I was scared of what waited for us there, even as a part of me felt certain that stepping into the unknown was the only way to chase something great.

WHERE TO LIVE

Once the decision was made, the next question was where to live. At CMU I would be paid a stipend of about $1,100 a month, with all tuition and fees covered. It was oddly satisfying in its own way — like getting paid to go to school again, the same feeling I'd had as an undergraduate on a football scholarship. (And this was long before NIL deals; back then a scholarship was the only form of "payment" you could count on.) Still, $1,100 a month wasn't enough to live on.

When we came to Pittsburgh, we did the usual rounds: apartments, small houses, whatever we could find. The apartments were out of reach financially, and the only houses we could consider were in the $50,000 range. Even that price stretched our budget. The catch was that we couldn't qualify unless Barb had a job she could put on the mortgage application. That's where a bit of luck and generosity came in. Our real estate agent's boyfriend pulled some strings and got Barb a job at Burlington Coat Factory in Monroeville. With that on paper, we qualified and bought a one-bedroom house for $49,000.

But Burlington was never the endgame. Barb kept hustling, scouring the Help Wanted ads in the newspaper and circling teaching jobs that looked promising. We drove around Pittsburgh with a map, trying to figure out where the districts actually were and whether they were good fits. During the day she worked at a local daycare, and in the evenings she clocked in at Burlington to keep us afloat. That daycare job turned into an unexpected break: one of the infants she cared for was the daughter of an attorney for Pittsburgh Public Schools. Impressed by Barb's care and reliability, the attorney got her an interview, which led to a teaching position in one of the city's underprivileged schools. It wasn't the perfect job, and at times it was tough, but it got Barb back into classrooms, working with kids — the thing she really loved.

Barb's Perspective — On Moving to Pittsburgh and Finding a Job

Moving to Pittsburgh was both challenging and fun. I had never been there, so learning the area felt overwhelming at first. But we were newly married, still in "college mode," and the adventure of exploring together made up for the stress. It felt easier once we started meeting people and building a circle of friends.

Finding a teaching job in Pittsburgh was hard. The profession was oversaturated, and without the Internet, figuring out where all the districts even were meant lots of maps, drives, and trial and error. I spent hours learning the geography of Pittsburgh's schools — which ones were in good areas, which were too far, which might be a fit.

It was frustrating, but it taught me persistence. Those early frustrations definitely nudged me toward grad school. I loved working with preschool kids, but I didn't want to teach at that level forever. Reading had always been a passion, so becoming a reading specialist felt like a natural next step.

PACKING UP AND MOVING

A few days after our honeymoon, Barb and I packed everything we owned into a 24-foot U-Haul. Hitched to the back was her 1990 silver

Chevy Nova, bouncing along as we pulled onto Interstate 64 heading east toward Louisville, KY and Cincinnati, OH.

My parents wanted to keep tabs on us during the trip, so they asked us to call whenever we had a chance. This was before the age of cell phones — so our options were limited to payphones and whatever coins we could scrounge. After Cincinnati, we took Interstate 71 to Columbus, OH and on that leg found a rural gas station, filled the U-Haul with gas, and went hunting for a payphone. I remember digging up a few quarters, sliding them into the slot, and calling Mom and Dad to let them know everything was going great. After Columbus, it was off to Wheeling, West Virginia, and finally into Pittsburgh, following I-70 and then I-79.

The drive took over nine hours in the U-Haul, so we didn't roll into town until late afternoon and immediately started unloading. That first night, we managed to move everything from the U-Haul into the first floor, spreading it around wherever it would fit.

Barb's Perspective — On Moving in the U-Haul

All I remember from the U-Haul move was having to go so slowly. With the full truck pulling the car, there was no way to go fast. It felt like crawling along, trying to get there safely.

Our possessions were modest. The centerpiece of our living room was a hand-me-down brown plaid, scratchy-fabric sofa and matching rocking recliner, probably from the late '70s or early '80s. The couch doubled as a pull-out bed, but it was massive — eight feet long and heavy as a tank. Getting it into the house turned into an ordeal. As twilight settled in, Barb and I wrestled the couch up the short stoop and through the front door. Except it didn't fit. It wedged itself in the doorway, completely stuck — me on the outside and Barb on the inside.

As luck would have it, a neighbor jogged by. I thought salvation had arrived. He introduced himself, glanced at the couch stuck halfway in our door, and said, "Looks like you've got that stuck." Then, to my disbelief, he jogged home without lending a hand. Not the best welcome to the neighborhood. Standing there next to a doorway too small for the life we were building, I realized how young we were — and how much we didn't know. But there was something liberating in that. We didn't have to know yet; we just had to keep moving.

Beyond the couch, we had my grandfather's full-size bed frame — a simple, sturdy piece he had used both on the farm and later in his city home. We had refinished it, and it gave us at least one connection to home. Soon after arriving, we bought a dining room set from a local furniture chain — a very '90s table with white tile inlays, light wood trim, four white chairs, and a hutch.

There wasn't much room for more. For the first few nights, we slept on a mattress in the dining room (using my old Boy Scout sleeping bag as a bed cover) because it was the only space with a window air conditioner. It was August and it was hot. Comfort came in the form of cool air, a mattress on the floor, and the feeling that we had just started something new.

The next day, we returned the U-Haul — an ordeal in itself. None of the roads in Pittsburgh seemed to follow a grid, and we had only one basic paper map to guide us. Winding our way through unfamiliar neighborhoods, we eventually managed to drop off the truck and begin the process of finding our footing in a city that was about to become home.

As it turned out, the next time I would make a long U-Haul drive like that — packing up everything and hauling it across states — was nearly 35 years later. This time, it wasn't for Barb and me starting out, but for moving our oldest son from his place in Manhattan down to Charlotte.

5430 WILKINS AVENUE

The house we bought in Pittsburgh was a narrow row house built around 1900, with just over a thousand square feet in total. It sat third from the end in a long stretch of 25 or 30 row houses that wrapped around the corner of Wilkins Avenue and Beeler Street — technically in Squirrel Hill, but right on the border of Shadyside.

Both neighborhoods were beautiful in their own way. Squirrel Hill was full of stately old homes, meticulously maintained, and anchored by a vibrant Jewish community — a cultural experience completely new to us, coming from the homogeneity of southern Indiana.

Shadyside, by contrast, had a more trendy, urban vibe, with shops, restaurants, and a bit of an edge. Our little row house was probably one of the only things in either neighborhood we could afford. But its location was unbeatable — less than a mile from Carnegie Mellon. I could walk or hop on my bike in the morning and be at class in minutes.

One of the things that stood out to me about Squirrel Hill's Jewish community was the Jewish Community Center — the JCC. In many ways, it was just like a YMCA — open to everyone, with flexible membership options including special rates for grad students. Barb and I joined soon after moving in, and it quickly became part of our routine.

I went there often to play basketball, which brought back good memories from high school and gave me much-needed exercise during those long, sedentary days of research. More than that, it gave us a way to feel more immersed in the neighborhood and connected to a community that was new to both of us.

Barb's Perspective — On Daily Life in Pittsburgh

I liked grocery shopping at the Giant Eagle on Murray Avenue in Squirrel Hill because it was small, and I've always preferred smaller stores. I also loved walking over to Shadyside, stopping in at the little card shop, and even getting my hair cut there. Coffee shops weren't "a thing" yet, so my routines were simpler — shopping, walking, and exploring the neighborhoods.

What stands out, though, was living in a community that was more diverse than anything either of us had grown up with. Coming from southern Indiana, this diversity was new to me, but I enjoyed it. It opened our eyes to different cultures and made Pittsburgh feel exciting and bigger than just the campus world we were part of.

Our next-door neighbor was an older, grandmotherly lady named Lotti, the kind of neighbor typical of her generation — always looking out for us (and sometimes a little nosy) but kind at heart. After we moved in, she was so appalled at the flimsy broom we bought to clear the sidewalk that she marched over and gave us one of hers. It was a small gesture, but one that made us feel welcome in a new city where we knew no one.

The house was only about fifteen feet wide and maybe thirty-five or forty feet deep on the first floor, and it was even smaller upstairs. From the outside, a simple stairway led up to the front door, and a large picture window gave the façade a little charm. Inside, our cat, Pumpkin, would sit on the back of a chair and stare out through that window, watching the cars and people go by.

We painted the front door a nice burgundy red, which gave the place a touch of warmth. There was also a noticeable flaw: a large crack in the concrete support beam that spanned between the basement and the first floor — one of those things you hoped wouldn't get worse while you were living there.

We tried planting flowers in the small front garden every spring, usually impatiens from a little neighborhood garden center just up Wilkins Avenue. We also planted an azalea bush that we thought would look great — until it was stolen. We added a planter box beneath the picture window, which brightened things up, and for a while the front of the house had some charm. (In a sad twist of fate, that very garden center was eventually sold, and part of the property became the Tree of Life Synagogue, where the horrific Pittsburgh shootings happened three decades later.)

Space was tight. The house had only six rooms: a decent-sized living room, a dining room we rarely used, and a tiny galley kitchen on the back where the floor sloped badly thanks to a long-settled foundation. The first floor also had two abandoned fireplaces — features that might have been cozy in the winter, given the inefficiency of the radiators, but they were not only inoperable; they almost certainly contained asbestos. We wisely never touched them.

Upstairs, there was one true bedroom with two small closets, a tiny bathroom, and a second "bedroom" that was so small we converted it into an office. That summer before moving in, my Dad built an extra closet in that little room because every bit of storage mattered.

The backyard was as quirky as the house itself. It was tiny and overgrown, and oddly there were no stairs from the main level down to it — even though there was a back deck. The deck was where we had to keep our garbage can, which meant every week we lugged the full can through the house and out the front door for pickup. It wasn't exactly sanitary.

As for mowing, we gave up on the old manual reel mower the previous owner left behind and always used a weed eater to keep the grass trimmed. Eventually, we cleaned it up and did some small plantings, which made it look decent, but since the only access was through a basement door, we hardly used it.

The basement itself was barely better than an underground dungeon. Rickety stairs led down to a low-ceilinged space where I had to duck to avoid hitting my head. The furnace worked well enough, but the washer and dryer were also down there. It wasn't a space you'd want to wash clothes in — dark, damp, and unwelcoming.

In the summer, the house was stifling. At first we had only one window air conditioner on the first floor, which made the upstairs unbearable. Eventually, we installed a small unit in the bedroom, though we were never sure the circuits could handle the extra load. The house was heated by steam radiators, which we had painted over, making them less efficient. One of the first projects we tackled was repainting the entire inside of the house, which at least made it feel fresh.

As you may be thinking, the house was a fixer-upper. The biggest remodel we undertook was the bathroom. We saved money by doing the demolition ourselves — chipping tile off the walls and floors, then hauling the debris away in five-gallon buckets to dump in Carnegie Mellon's dumpsters late at night.

When the room was finally cleared, a contractor handled the remodel, except for the cast-iron tub. Barb re-enameled and painted it, turning it into one of the nicest features in the house. For weeks during that remodel, we drove to CMU each evening to shower in campus facilities and timed our bathroom breaks around the fact that our only bathroom was gutted.

Later, we tackled our bedroom, adding a chair rail, wallpaper, and carpet. It turned out looking nice, even if my carpentry skills left plenty of rough edges.

Despite the limitations, the little row house became our first real home. We lived there seven years before selling it and moving to the suburbs, and fortunately we recovered the money we had put into it. Even today, whenever we're back in the city, we make a point to drive by. For a while it looked fine, but over the years it's become more and more dilapidated — a little sad to see.

And then there were the quirky neighborhood customs. One of the strangest came in the winter: when you dug your car out of a snow-covered street parking spot, you "reserved" it by leaving something in the space — often a kitchen chair. At first it seemed bizarre to see chairs sitting in the middle of the street, but eventually we understood. If you did the hard work of digging out your spot, you didn't want someone else taking it. It was just one of those small cultural lessons that made Pittsburgh unique.

Barb's Perspective — On Our First Home

> *It was fun decorating our first home together. It was hot, but since I didn't grow up with air conditioning, it didn't bother me too much. I loved our ritual of walking around the neighborhood in the evenings and dreaming about someday living in one of the bigger houses nearby. Those early months were when I first realized I could build a life anywhere, as long as Todd and I were building it together.*

THE ROBOTICS INSTITUTE

While our home life was slowly settling, CMU was another story entirely. Nothing felt familiar there — not the buildings, not the people, not even the language. Walking into Wean Hall for the first time felt like stepping into a different world and I didn't grasp just how remarkable the place I was entering would prove to be.

Carnegie Mellon's Robotics Institute was born in 1979, conceived at a time when robotics was still a dream, not a discipline. It was launched with bold ambition — envisioned by President Dick Cyert, Deans Dan Berg and Angel Jordan, and legendary pioneer Raj Reddy — with early funding secured through a five-year, $5 million commitment from Westinghouse and the Office of Naval Research. The goal was audacious: build the world's first and foremost center to pursue "thinking robots," marrying fundamental science with real-world application, and become a hub for innovation driving everything from manufacturing to space, medicine, and autonomous vehicles.

By the time I arrived, it had matured into one of the largest and most respected robotics research institutions on the planet — setting the stage for every experiment, prototype, and late-night line of code that would follow.

It's here that the first Robotics Ph.D. program in the world was founded (officially in 1988). The inaugural "unofficial" class began around 1989, and the first official class — roughly thirteen of us — arrived on campus in the fall of 1990. That group was small, diverse, and full of energy, the kind of people who had to believe in possibility more than precedent. I take special pride in knowing I was part of that very first class and ultimately became the tenth person in the world to earn a Ph.D. in Robotics.

Over the years since, the Robotics Institute (RI) has grown far beyond what anyone could have imagined. Today it boasts around 71 faculty members dedicated solely to robotics research, with dozens more courtesy, adjunct, and affiliate faculty across related fields like AI, computer vision, and human-robot interaction. The labs and groups have proliferated, covering everything from soft robotics to multi-agent systems, legged locomotion, autonomous vehicles, surgical and assistive robotics, and perception under extreme sensing conditions. Funding and facilities have scaled up too, with annual research expenditures in the tens of millions and a new Robotics Innovation Center underway that will add 150,000 square feet of flexible lab and testing space.

I was fortunate to be there near the beginning — close enough to feel the raw energy of a place just finding its stride, yet early enough that each of us still helped write its story. Watching RI grow from those days to its position now as the undisputed global leader in robotics has been one of the most satisfying arcs of my professional life.

Carnegie Mellon's Robotics Institute at that time wasn't a single building so much as a constellation of people scattered across the School of Computer Science and the School of Electrical Engineering. Offices and labs were tucked into corners all over campus, but my world centered on Wean Hall — an 11-story concrete block in the Brutalist style, more fortress than academic home. (Brutalist style — which is exactly as cheerful as it sounds.) It was the hub of the School of Computer Science, housing the SCS library, the main computer operations center, and a warren of labs and offices. Many of those offices had no windows — mine included — with the coveted corner and window spots reserved for faculty and senior grad students. Inside, the atmosphere was intense, whiteboards crammed with equations, the hum of workstations, grad students hunched over Sun terminals, and pockets of researchers grouped by specialty — theory and compilers in one area, ML and AI in another, robotics scattered wherever we could find space.

For us in the Robotics Institute, space mattered. Unlike the theorists, we needed rooms big enough to roll in vehicles, set up sensors, and wire together hardware. That made us the hands-on crowd, and I thought we were sometimes dismissed as "almost CS" by our more theoretical peers. But we had Raj Reddy on our side, and that gave us credibility. To me, Wean Hall felt intimidating and sterile at first — all sharp lines and raw concrete, a place built for utility rather than comfort. Yet behind that harsh exterior were the people and ideas that would completely reshape my life.

I still remember walking into my very first office — a windowless, concrete, former equipment room that I shared with two foreign students and another from a big-time school. I dropped my bag, looked around at the bare walls, and thought to myself, "Maybe this wasn't such a good idea." Coming from Indiana State, where the campus was modest and familiar, Wean Hall felt enormous and overwhelming. My new office mates came from all over the world, and at first, I could barely follow their accents. For a kid from Huntingburg, Indiana, it was disorienting.

What I didn't realize then was that the same building I once doubted I belonged in would soon become the center of my world — a place where I'd work deep into the night, wrestle with problems, and begin to prove to myself that this was exactly where I was meant to be.

THE HALLWAY

The ground floor hallway of Wean Hall became the axis of my life. On the left were the faculty offices, along with senior grad students and staff members, their windows looking out toward the sunlight. On the right were the labs. Most of this floor housed researchers from the Vision and Autonomous Systems Center (VASC), where the Navlab group worked. The actual Navlab 1 — a big blue panel van — was kept nearby at the Field Robotics Center (FRC).

The hallway offices and labs were the nerve center of VASC, and inside these walls most of the early breakthroughs at CMU in computer vision and mobile robotics were happening. Takeo Kanade was the VASC head when I arrived, and acronyms were everywhere. (As you'll notice throughout, acronyms in academia are a kind of techno-currency: buildings, labs, projects, software — everything needed one. They had to be both descriptive and pronounceable, and sometimes the act of naming felt as important as the work itself.)

Up to now I've talked about the space — the concrete walls, the desks, the Coke cans — but the real story was the people. That's where CMU's atmosphere really came alive.

The first office on the left belonged to Chuck Thorpe. I met him right after completing the Robotics "immigration course" — CMU's intense orientation for new grad students. It was equal parts boot camp and gauntlet. We learned the essentials of surviving at CMU: how the School of Computer Science and the Robotics Institute worked, the maze of administrative hurdles, and — most importantly — hearing presentations and seeing demos from the faculty who were looking for new students.

For most of my classmates, the talks were a chance to connect with advisors in fields they were already familiar with, or to continue research they had begun as undergraduates. For me, it was daunting. Nearly every talk went completely over my head. I had no background, no contacts, and very little idea what the professors were talking about. My rankings came down to a simple test: *what seemed cool?* It wasn't exactly speed dating, but that's what it felt like — a few days to make choices that would shape your entire Ph.D. You were on your own to figure it out, and then the program "matched" you with an advisor.

By the end, I had narrowed my list to a couple of professors, all working in mobile robotics. Chuck was at the top. When I finally sat across from him in his office, I didn't even know what to ask; I wasn't fluent in academic jargon yet. But Chuck was patient. He listened, asked me questions, and never made me feel stupid. He must have seen something in me — or maybe he was just desperate — because he agreed to be my advisor. That decision changed everything.

Farther down the hall was the Leg Lab, run by Marc Raibert, where the first hopping robots were being built. They looked more like primitive animals than mechanical contraptions. Raibert would go on to found Boston Dynamics, whose robots decades later would both amaze and sometimes terrify people — not because they failed, but because they moved with an unsettling realism. (In a twist I couldn't have imagined back then, my daughter now works in the very same building that currently houses Boston Dynamics.) In 2006, one of those robots and one of mine were inducted into the Robot Hall of Fame in the same class. Back then, though, it just felt like being surrounded by brilliance.

Next was the computer lab. It was the real hive of activity, mostly for junior grad students. Harsh fluorescent lights buzzed above, and there were no windows. Soda cans and fast-food wrappers littered the desks, trash bins overflowed, and the air sometimes carried the smell of students who clearly hadn't slept — or showered — in days. Rows of Sun workstations filled the room — some of the older, bulky models alongside the newer Sun "pizza box" designs that housed the latest SPARC processors.

The best machines were reserved for senior students who didn't already have one tucked into their offices, and a few had special displays attached and were used by the computer vision researchers. Usually there were two or three of us in there, but sometimes as many as fifteen, crammed shoulder to shoulder, typing away late into the night. For newcomers like me, it was a cluster of shared machines; for others, it was where bigger jobs got done when their office setups weren't enough. The lab itself had low-cut carpet, reasonably comfortable chairs, and a high ceiling threaded with network cables running every which way.

What made the lab special, though, was the people. In general, the folks there could answer almost any question you had — whether it was about C coding, the CMU network, or the nuts and bolts of basic robotics. It was like a mosh pit of information: noisy, chaotic, but full of energy. You learned just by being present — by asking questions, overhearing answers, or helping someone else, and in the process picking up what they were working on. It wasn't a place for quiet genius; it was a place where collaboration thrived.

Nothing like that crowded, noisy lab had ever been part of my experience before. Coming from Indiana State, the contrast couldn't have been sharper. High school and ISU had given me good teachers and mentors like Mr. Menke and Dr. Cockrell, but there was nothing like this — no constant cross-pollination of ideas, no community where you could literally learn just by hanging around. It wasn't a revelation so much as a new kind of air to breathe, the kind of environment I hadn't known I was missing until I stepped into it.

Behind the computer lab was a little back room with a fridge stocked with Cokes — fifty cents apiece, on the honor system, with an empty coffee can for the money. Those Cokes fueled more debugging sessions than I can count, and on some days they substituted for meals — which, in my twenties, I could get away with. Later I had to switch to Diet. Some students stashed food in there too, turning it into a kind of communal fridge for grad-student survival. For me, lunch was usually takeout, and Coke covered mornings and late nights before dinner. As gross as it sounds, a Coke in the morning did a fine job stripping the gunk off your teeth after a night of coding. (Not exactly dentist-approved, but it worked.) A cold can of Coke was a small luxury in a place otherwise built on stress. Eventually, my office moved into that hallway, and I shared it with Omead Amidi.

OMEAD AMIDI

Omead was patient, brilliant, and generous. His parents had emigrated from Iran just before the 1979 revolution, and he always insisted on describing himself as Persian, not Iranian. For me — a small-town kid who had barely left Indiana — that distinction was fascinating. We had long conversations about his heritage, politics in the Middle East, and what it meant to grow up between cultures.

He was also a tremendous researcher — the exact kind who thrived at the RI: technically brilliant, but also driven to build things. In the Navlab group and across VASC, Omead was the godfather of hardware and control theory. He pioneered early fuzzy-logic experiments with autonomous helicopters — not the little hobby drones you see today, but quarter-scale machines ten feet long, powered by big engines — capable of real applications. One of the first use cases was crop dusting. His vision-based control work branched into other surprising areas too, including high-speed beer inspection. The Kirin Beer Company in Japan hired VASC, Takeo, and Omead to build a system that could automatically check bottles racing past on the line at industrial speed.

For me, Omead was a lifeline. He showed me how to navigate the strange world of robotics research, how to approach problems systematically, and how to keep a sense of humor even when things broke for the tenth time that day. He knew every quirk of the Navlab

1, which was nothing like a typical car you just climb into and drive. You had to top off gas, start an auxiliary engine, flip a dozen power switches, manage Internet connections depending on the test, check the hydraulic transmission, and always have a safety driver. Without him, I never would have survived the early years.

Later in life, I served as interim CEO of his company, Skeyes Unlimited. I'm not sure I was a great CEO, but I did help them land some work, so at least I paid my way. To this day, Omead is the best control-systems engineer I've ever known — and also one of the kindest, most patient people you'll ever meet. If you ask him, he'll deny being patient and maybe even kind. But don't let him fool you. He is.

A PRESENCE IN THE BACKGROUND: RED WHITTAKER

One of the towering figures at the Robotics Institute in those years was William "Red" Whittaker. I wasn't in his lab, and I never did a project directly under him, but his presence was unmistakable. He was one of the first names I heard when I arrived at CMU, and even from a distance, he was hard to miss.

Red grew up in Hollidaysburg, Pennsylvania — a small town not unlike my own Huntingburg, Indiana. His father sold explosives, his mother was a chemist, and he carried that mix of grit and technical sharpness into everything he later did. He studied civil engineering at Princeton, interrupted his studies to serve in the Marines, then came to Carnegie Mellon for graduate school, earning his Ph.D. in 1979. Almost immediately, he was thrust into one of the most urgent technical crises of the time: the near-meltdown at Three Mile Island. With a budget of just $1.5 million, Red and his team built robots to inspect and repair the reactor's damaged basement. That project became the seed of what would eventually grow into the Field Robotics Center.

Red had a striking and unusual way of speaking — with pauses that felt like he was finishing the rest of the sentence in his head. They didn't come across as awkward, just... different. He could be demanding, even harsh. At first, I thought it was too much, the kind of drill-sergeant mentality that didn't fit an academic environment, even one like CMU RI. But as I got older, and especially after seeing what he went on to accomplish, my perspective changed. He was, in his own way, a master motivator. He dared people to think bigger, to attempt the impossible, and by sheer force of will, convinced them it could be done and that they could do it.

His vision was audacious. The Field Robotics Center under his leadership developed machines to inspect nuclear reactors, explore volcanoes, search for meteorites in Antarctica, and map collapsed coal mines in Pennsylvania. His teams pushed robots into places no human could safely go — the Chernobyl reactor, the Atacama Desert, and even simulated Martian terrain. Later, he became one of the leading figures in the DARPA Grand Challenges, the competitions that catalyzed the modern self-driving car industry. His vehicles *Sandstorm* and *Highlander* placed second and third in 2005, and in 2007 his "Tartan Racing" team took first place with *Boss*, winning the $2 million prize.

Even the moon wasn't too far. Red led CMU's entry in the Google Lunar X Prize, pushing for a robotic landing on a shoestring budget, and his later MoonRanger project extended that dream under NASA contracts.

In those early days at CMU, I didn't fully appreciate him. To me, Red was this larger-than-life figure who seemed too harsh, too single-minded. But now I see him differently. He was a visionary leader — the good ol' boy version of Elon Musk — unafraid to sound crazy, unafraid to fail, because he knew his drive would pull others along with him. In many ways, I've come to realize I'm more like him than I ever would have admitted back then. We were both small-town kids who believed bold ideas, backed by hard work, could change the world. And I'm proud of that.

FINDING BALANCE ON THE FIELD

Another peer who made a big difference in those early years was Kevin Lynch. Kevin was a true local — he grew up in Monroeville, a suburb of Pittsburgh, where his dad even served as mayor for a time. He went on to Princeton, where he played defensive line on the football team, before coming back home for graduate school at CMU. That meant we shared something rare in the Robotics Institute: the identity of being both athletes and academics.

All of our other classmates came straight from elite technical programs and had never strapped on a helmet. Although Kevin went to Princeton and I to Indiana State, we both understood the grind of two-a-days, the camaraderie of locker rooms, and the toughness it took to keep pushing when your body hurt. That common ground gave us an instant connection. Even though our research never really overlapped, we hung out frequently.

It also led us to some unlikely adventures. One of the quirkiest was when Kevin and I ended up coaching a Japanese women's flag football team. Takeo had arranged it — even though he knew almost nothing about football — and, well, you didn't say no to Takeo. A dozen women had come to CMU from Japan, at least in part to learn football, and suddenly we were their coaches. It was about as far from robotics as you could get, yet it felt natural. Coaching drew on the same skills we'd learned in sports and in the lab: patience, clarity, teamwork, and discipline. Plus, it was fun. In the middle of late nights coding and endless debugging, stepping onto a practice field to run flag football drills reminded me I was still that kid from Indiana who loved the game.

Kevin also introduced me to the School of Computer Science intramural flag football team, hilariously named the *NP Completions*. For computer science folks, the name was a pun on the NP-complete class of problems — basically the hardest kind — so calling ourselves the *NP Completions* was both geeky and tongue-in-cheek. The team already had a quarterback, Dave Kosbie (Koz) — a high-strung, brilliant guy from Harvard. He was solid, if not spectacular. At first, I tried to be generous and line up at wide receiver or somewhere else, but it quickly became obvious I belonged under center. Compared to Koz, I was playing at an NFL level. (Don't be mad at me, Koz!)

As I remember it, once I took over as quarterback, we started winning — a lot. We even won the intramural championship, beating undergraduate teams along the way, and took home the coveted championship T-shirts as proof. To his credit, Dave was fine with the switch. I'm pretty sure we were the best graduate School of Computer Science intramural football team in the history of the world. And for all the talk about MIT, Stanford, and Berkeley being our peers in research, let's be honest — we were better at robotics, and we sure as hell had the better intramural football team.

And then there was Shakey's Sluggers, a co-ed softball team made up of grad students, spouses, and friends. Barb was our all-time pitcher, and she was a good one — tough, competitive, and steady on the mound. Unfortunately, she had a habit of stopping line drives with her shins, and for years after CMU she carried a permanent bruise as proof. I worried about her, but she could handle herself just fine, and sometimes she even got mad at me when I cautioned her to "be ready."

Barb's Perspective — On the Softball Team

I have fond memories of our intramural softball team. It was a fun, social event, and sports have always been an enjoyable part of our lives. It was wonderful to do it together with friends, and sometimes we'd head to the bars afterward to keep the camaraderie going.

We played our games on the CMU football field, which sat inside a small stadium. Right field was a short "porch," thanks to the bleachers crowding in, and I still had enough pop in my bat to launch home runs deep into those stands. It felt a little like being a pro, and I'll admit I took pride in it. Most of the academic folks around weren't great athletes, and at least out there on the field, I was better than most of them — even if in the classroom it often felt the other way around.

Finally, one of the more eccentric but unforgettable traditions at VASC in the early '90s was our golf outings. From 1991 through 1995, we held what became known as the VASC Golf Tour. These weren't just casual rounds. They came with full write-ups, leaderboard drama, and an ever-expanding cast of nicknames. Takeo was "The Golden Bear," Dean became "The White Shark," Chuck showed up as "E-Cheese," and there was even a "Big Daddy" and "Monster Driver" in the mix. It was absurd. It was glorious.

Barb and I played, too. We didn't have nicknames, but we showed up, made it through 18 holes, and occasionally found ourselves part of the chaos. One year I ended up at the center of a completely fabricated scandal — accused of inflating my score so my advisor could take home gross champion, supposedly in exchange for a fast thesis sign-off and a million bucks. That was the tone of it: completely unserious, and somehow deeply memorable.

The golf tour wasn't official, and no one took the scoring seriously — except for Takeo and Dean. But it gave us a way to connect outside the lab, to roast each other a little, and to build the kind of camaraderie you don't get just standing around a whiteboard. Some of the best stories from my CMU years came not from conferences or experiments, but from those humid afternoons out on the fairway, with a group of brilliant people who also happened to be terrible at golf.

Sports, whether coaching, the NP Completions, Shakey's Sluggers, or playing golf poorly gave me balance. They reminded me that while I might have felt out of place in Wean Hall at first, there was room at CMU for people like me — someone who could love both the game and the code.

FOOD

In Wean Hall, it was everywhere and never lasted long. Just outside our hallway, every day like clockwork, a man from a local Chinese restaurant set up a lunch counter. He brought noodles, rice, Kung Pao chicken, and other dishes. The smell drifted through the building, and for a kid raised on Midwestern casseroles, the aroma was exotic. And it was cheap: three dollars for a heaping plate. Nearly every day, I ate there. It became one of those small rituals that grounded me during the chaos.

If Coke was the drink of choice, pizza was the food of choice. Pizza could draw a massive crowd to an otherwise painfully boring lecture. It was the centerpiece of nearly every social event: box after box stacked high on folding tables. You could even tell the importance of the occasion by the kind of pizza. If it was a big deal, it came from Mineo's or another "prestige" shop. If it was just a routine gathering, it came from whatever local place was closest.

There were also unwritten rules about pizza. You never took it home. You ate it there or left it for the next person. Even if there was one slice remaining, you didn't squirrel it away. It was survival of the fastest to the table, but it was also a form of shared understanding. Everyone knew the situation, and nobody hoarded — we were all in the same boat.

On the home front, food looked a little different. The place Barb and I ate at the most was Taco Bell. We ordered the same thing nearly every time: ground beef or black bean soft tacos and those little bags of nacho chips with cups of bright orange nacho cheese. I'm sure it was terrible for us, but we loved it. And I hate to say it, but even when Barb was pregnant with Emma, Taco Bell was still a mainstay in our diet. To be fair, Barb generally ate better than that, but we both ate more Taco Bell in those years than I care to remember.

Food, in all its forms — Chinese lunches in the hallway, Coke-driven late nights, pizza-fueled lectures, and Taco Bell dinners — became part of the fabric of grad school. It wasn't just fuel. It was culture, survival, and sometimes even comfort. Sharing food was another way I found belonging in a place that could otherwise feel cold and overwhelming.

SPORTS IN THE "CITY OF CHAMPIONS"

Life outside of Wean Hall wasn't all debugging and Coke. Pittsburgh itself had its own rhythms, and one of the biggest was sports. The Steelers' four Super Bowl titles in the 1970s defined the city's identity, and in 1982 Howard Cosell famously dubbed it the 'City of Champions.' The Pirates' 1979 World Series win only added to that swagger. The city lived and breathed its teams, and Barb and I quickly got swept up in it — even in sports we barely understood at first.

When we moved in 1990, the Penguins were led by Mario Lemieux. Neither of us had grown up with hockey — there wasn't a team within 200 miles of Huntingburg — but in Pittsburgh, it was unavoidable. We learned the rules as we went, and before long, we were true fans. It was so cool watching them beat the Minnesota North Stars 8-0 in Game 6 of the 1991 Stanley Cup Finals — Lemieux scored in the first 30 seconds, setting the tone for their first championship. They repeated the feat in 1992, though the years after brought tougher times. One moment etched in my mind was Kevin Stevens' devastating injury in the corner during the '93 playoffs — a reminder that the sport we'd come to love could be brutal.

The Steelers, of course, were the heartbeat of the city. By 1996, it had been a decade and a half since their dynasty years, and Bill Cowher had taken over from Chuck Noll. With Neil O'Donnell at quarterback and a defense stacked with talent, they clawed their way back to the Super Bowl against the Cowboys. From my perspective as a former quarterback, the pivotal interception wasn't O'Donnell's fault — the receiver ran the wrong route — but it sealed the loss. That was the last shot until the Roethlisberger years revived the franchise — years later I'd be there in person at Super Bowl XL in Detroit, and then with my whole family in Tampa and Dallas, watching the next great Steelers run unfold.

Even baseball was fun then. The Pirates were competitive, with stars like Bobby Bonilla, Barry Bonds, and Andy Van Slyke roaming the outfield. Barb and I even bought a partial season-ticket plan — about 20 games in left field. Jim Leyland, their chain-smoking manager, had previously coached the Evansville (IN) Triplets, so I'd heard of him long before Pittsburgh. The Pirates made the playoffs three straight years but suffered heartbreaking exits, none worse than Sid Bream — a former Pirate — sliding home safely against them in the '92 NLCS because Barry Bonds couldn't throw him out.

It was a great time to be in Pittsburgh. Pittsburgh's sports culture gave us something outside the lab to get excited about — a way to feel tied into the city we now called home.

LIFE OUTSIDE THE LAB

Barb and I didn't spend much time in bars or clubs like some of the other grad students. We didn't have the money for it, and honestly, it just wasn't us. Our social life revolved around other married grad students — dinners in tiny apartments, game nights that went too late, and small trips like rafting on the Youghiogheny River. Nothing flashy, nothing wild. Just shared food, shared stories, and a chance to step outside the grind for a few hours. It was never really about where we went; it was about being around people who understood what it meant to be young, married, and trying to survive grad school together.

And that's where one rafting story comes in. We went on those trips nearly every year. They weren't big-water, adrenaline-junkie rapids by any means. They felt big to us, sure, but they were completely safe — no real experience needed, no guide sitting on the back yelling commands. Just a bunch of grad students in a couple of rafts, drifting down a river that made us feel a little tougher than we actually were.

On this particular trip, Barb and I were sharing a boat with a couple of the others and Kevin Lynch — who you might remember had played defensive line at Princeton. Big guy. Probably 6'1", 250. As we hit one of the faster sections, the raft started wobbling, and before any of us could grab him, Kevin went right over the side.

He was never in any danger, but getting him back in turned into its own comedy. We tried to haul him up from the water, but between his size, our lack of coordination, and the raft bouncing around, it just wasn't happening. After a minute of struggling and laughing, we finally gave up, paddled to the bank, and let Kevin climb back in from solid ground. Then we pushed off like nothing had happened, still laughing about it the rest of the way down the river.

NOT MY BEST MOMENT

One moment from those years that still makes me cringe came on our first wedding anniversary in 1991. Barb had gone out of her way to make it special. She made reservations at Christopher's on Mt. Washington — one of the nicest restaurants in Pittsburgh, known for its seafood. It was the kind of place that made you feel like you were part of the real world, not just two overworked kids grinding through grad school.

What I didn't recognize then, but can see clearly now, is just how tightly wound I was at that point. I had just completed my first year at CMU. I was about halfway through my classes and finally starting to feel like maybe I belonged there, but at the same time I knew how much I still didn't know and how much further I had to go. I still had courses to finish. I still needed to come up with an actual thesis idea and defend a proposal. And Barb and I were making it, but we didn't have much money. Every dollar mattered.

I had also recently been home and seen how some of our friends were moving on with their lives — real jobs, real paychecks, a sense of stability we didn't have. All of that built up inside me in ways I didn't fully understand then. So when I walked in the door that night, tired and stressed, worried about money, and Barb told me she had made reservations at a nice restaurant — it all hit at once. Instead of being grateful, I snapped at her for spending too much. I overreacted. I was a jerk. I think I even made her cry.

Looking back, it wasn't really about the restaurant. It was every insecurity, every pressure point of that year — money, school, expectations, fear of not measuring up — all getting wrapped into one moment. The idea of going out and spending money just pushed me over the edge.

We went anyway, and of course it turned into a wonderful night. But that moment stuck with me. It's one of those small, important lessons: don't let stress make you blind to the good someone is trying to do for you. Barb wasn't being careless. She was trying to remind us that, even in the middle of all the chaos, we were allowed one night to feel like more than broke grad students. And she was right.

IMPOSTER SYNDROME

But inside CMU, the biggest challenge wasn't food or even the environment. It was inside me.

For much of my life up to that point, I'd been a big fish in a small pond. In small-town southern Indiana, I'd found success in sports and academics, and that brought me plenty of attention. At Indiana State, I thrived in the classroom, and being on the football team gave me extra visibility with professors and administrators. But CMU was different. Suddenly I wasn't the big fish anymore — I was a goldfish dropped into the Pacific. It was humbling. It was scary. But it was also exciting. For the first time, I had the chance to truly measure myself, to find out what kind of person I could be and what I could achieve.

Many of my classmates had graduated from MIT, Stanford, or Berkeley. They spoke the language of research fluently. I came from Indiana State. I hadn't programmed in Lisp or C. My only programming experience since high school was toggling switches on an 8-bit microcomputer in assembly language. When I told people where I was from, nobody ever said anything unkind, but you could see the flicker in their expression: *Indiana State? Really? How did he end up here?*

So I started from the bottom. I taught myself the tools everyone else already seemed fluent in — Lisp first, then C, the lower-level workhorse that most of the research code was written in. (Lisp was the language with all the parentheses. You either loved them or drowned in them. I mostly drowned in them.) I read manuals late into the night. I learned Unix, the command line, compilers, makefiles, and shell scripts. It was overwhelming, and there were nights I felt like I'd never catch up. But the culture at CMU, competitive as it was, was also generous. Whenever I asked a question, someone helped. Even the smartest students would sit down and walk me through problems. That ethos — work hard, ask for help, then help others — kept me afloat.

It was funny how much of this mapped onto football. In football, especially in the offseason, you grind for months with no games in sight. You lift weights, run drills, and push through the monotony, trusting that when the season comes, the payoff will be there. Grad school felt the same way. For months on end, it seemed like I was working without making any progress, never quite getting to the "game." But football had already taught me what to do in those moments: keep showing up, keep putting in the work, and trust that when it mattered, I'd be ready.

In that grind, I slowly began to realize I did belong — not because I started ahead, but because I kept showing up.

FIRST PROJECTS AND SMALL WINS

The first project I remember being assigned was programming an A* planner. A* was a search algorithm originally developed in the late

'60s for Shakey the robot at SRI. (As you may recall, Shakey also lent its name to our grad-student softball team, Shakey's Sluggers. Yes, I know. Very geeky.) A* mattered because it became the backbone of path planning for mobile robots — the logic that allowed a machine to find the most efficient route from where it was to where it needed to go.

The assignment sounded deceptively simple: write the algorithm in C, make it work, and build a graphical interface in X Windows to show the path. X was one of the earliest and most powerful graphical windowing systems for Unix, providing a front end for drawing and interaction on screen. (And no — not *that* X, the one that replaced Twitter decades later. This one was older, crankier, and a whole lot less interested in your hot takes.) For me, though, this project was like trying to read a foreign language with no dictionary.

It nearly sank me. I had to learn recursion, memory management, and graphics programming all at once. Recursion was especially brutal — if you got it wrong, the program went spiraling into Neverland, chewing up memory until it crashed in spectacular fashion. Just getting the program to compile was its own mountain to climb. Syntax errors when you didn't even know what "syntax" meant. Linker errors, missing libraries, misconfigured compile paths, and the endless trial of typing cc only to see another cryptic error message. Then there were optimization flags — do I use -O2, -O3, or -O4? Each one seemed to promise a faster program but more often left me confused.

Programming graphical applications on Unix back then was no easier. Everything started with the X Window System (X11), but you didn't just write simple drawing commands — you had to go through raw Xlib, this low-level, C API built out of opaque handles, pointers, and layers of indirection I'd never seen before. On top of that, any real application used Xt or Motif, which added their own ideas of widgets, inheritance, and resource files. Using X felt like being handed a professional graphics stack designed by people who hated simplicity. Even learning the vocabulary took weeks.

After countless late nights, I finally got it running. On my screen appeared the results of the search: a pathway drawn from start to finish, the "least-cost" path winding its way through the maze. It wasn't actually a robot moving, but it was my first real proof that the code worked. It may not have been as exciting as driving a car, but it was a start — and more importantly, it proved I could get better. I probably spent as much time trying to get this simple little program and interface working as I would later spend on the first drafts of much more complicated algorithms that became the foundation for autonomous cars.

That project mattered not for its academic weight but because it threw me straight into the deep end. I had to learn the tools of the trade: UNIX, compilers, makefiles, debuggers, scripts. I had to learn how to survive as a programmer, and those early lessons became the price of admission for the research that was coming.

THE COHORT

Our Ph.D. class was a mix of brilliance and fragility. True grit was rare — the kind forged outside of the classroom — and it was the one edge I carried with me. Some students were there to prove how smart they were. Others quietly helped whenever they could. And some of the brightest students never made it.

One classmate was brilliant but social — too social for the program's workload — and decided the five-year slog wasn't worth it. Another ran into money problems and had to leave early. Two women entered with stellar credentials. One was probably the smartest in our class, but she could never settle on a thesis topic and eventually left. The other was sharp too, but she carried herself with a certain arrogance. One day in class a professor dressed her down in front of everyone. It was brutal — she literally cried — and she never recovered.

Watching that unfold was a stark reminder that perfection — or pride — could become the enemy of success.

For those of us who stayed, the Robotics Institute became a crucible. You were free to chase your own ideas as long as each had some plausible tie back to your project research. And the resources were there — not unlimited, but if you had a reasonable case for why you needed them, you could get access. In that way it reminded me of my hometown. Back there, in a small town built on manufacturing, if you worked hard and pulled your weight, people pitched in to help. At CMU, it wasn't factory workers — it was world-class roboticists and coders. But the ethic was the same: work hard, do your part, and people had your back.

Watching classmates leave — some of the brightest among us — also left its mark. It showed me that brilliance alone wasn't enough. In the end, persistence mattered just as much. Sometimes the biggest difference between those who finished and those who didn't was simply sticking it out.

THE CASE OF THE ECHO NAMES

Every place has its quirks, and CMU was no exception. One of the strangest anomalies I ever ran across in my life happened there: two students in the program, both with first and last names that were exactly the same. The first was Herman Herman. If I remember correctly, he came from Indonesia, where family names aren't always used. When he arrived in the U.S., his passport simply read "Herman." Immigration officials told him he needed both a first and a last name, so he shrugged and said, "Okay, Herman." And that's how he became Herman Herman, a name he kept through grad school and into what turned out to be a very successful career. The second was Martin Martin, a Canadian student. In his case, it wasn't a bureaucratic quirk but a genuine family name passed down. Different backstory, same result: two guys with Echo Names sitting in the Robotics Institute. It was one of those little oddities that stuck with me — something you'd never expect but that gave the place its own flavor.

REFLECTIONS

Those first CMU years taught me how place, people, and just showing up day after day braid together. Wean Hall may have looked like a concrete fortress, but inside it was a living network — VASC lining the hallway, mentors who opened doors, and peers who answered questions at 1 a.m. Just across the drive, Navlab operated out of a garage bay, a proving ground where ideas became machines. None of it was glamorous, but together it was exactly the kind of arena where you either found your footing or you didn't.

I had come to CMU feeling out of place, but after a few years, persistence and small wins proved I belonged. Progress stacked quietly, one project at a time, until it became something I could point to with pride.

The culture mattered. CMU could be sharp-edged and competitive, but it was also generous: ask a good question, get a good answer; do the work, earn the trust. What really stitched the human side together were the shared rituals — food grabbed between experiments, late nights in the lab, intramural football, Shakey's Sluggers with Barb on the mound, and simple dinners or rafting with friends.

I also saw how fragile the path could be. Brilliant people left — for money, for pride, for life. It taught me that excellence isn't just intelligence; it's accountability, patience, and showing up again tomorrow. Winning matters, but not because it's flashy — because it signals preparation, toughness, and follow-through.

If there's a through-line from Huntingburg to Wean Hall, it's this: success wasn't about being the smartest in the room. It was about refusing to fade. That foundation carried me forward. Next came the real test: the grind of coursework and the mentors who shaped how I thought, not just what I did.

CLASSES AND MENTORS
LAYING THE FOUNDATION

Graduate school at Carnegie Mellon wasn't just about surviving the lab. The coursework was its own forge, shaping me in ways I hadn't anticipated. Coming from Indiana State, I thought I was a solid student. But the classes at CMU — taught by some of the most brilliant minds I'd ever encountered — showed me what real intellectual rigor looked like. They didn't just test what I knew; they reshaped how I thought. And somewhere in the middle of all that pressure, a quiet, unsettling realization crept in: maybe I actually could do this. Not confidently, not cleanly — just maybe. It was a terrifying kind of hope. I could see enough of the path to imagine finishing, but not enough to know whether I was truly capable of it.

That foundation — equal parts theory, systems, and problem-solving — became the toolkit I would later lean on for my thesis work and for projects like No Hands Across America and the Automated Highway System demo. But at the time, that toolkit felt like it came with its own weight: the fear of failing on one side, and the fear of succeeding — and having to live up to it — on the other. It was the start of a leap I wasn't yet sure I could make.

THE CLASSES THAT SHAPED ME

Three courses stand out in my memory — Manipulation, Perception, and Mobile Robotics. Together they laid the intellectual foundation for everything I would later do with autonomous vehicles. The funny thing is, these weren't graduate seminars — they were actually 300-level undergraduate courses. Every new robotics Ph.D. student had to take them unless you could get a waiver. I couldn't, of course. The only robotics class I'd ever taken before CMU was a single, very basic course back at Indiana State.

So there I was, a brand-new grad student, sitting in with undergrads and thinking to myself: *If this is what they expect out of one class, what must it be like to take an entire load at this level?* It was humbling. CMU wasn't just an incredible grind for graduate students; it was every bit as tough for the undergraduates. In a way, seeing that up close made me respect the culture even more. The bar was high across the board.

MANIPULATION, WITH MATT MASON

Matt Mason looked every bit the professor — graying beard, glasses, and a slightly rumpled but unmistakably authoritative presence. He was taller and thin, with a rugged but lean look — almost like a geek crossed with a cowboy — and I remember he was partial to plaid shirts. What surprised me later, though, was that he was only in his late thirties. He was already an established giant in robotic manipulation and would eventually go on to head the

Robotics Institute. Outside the classroom, he'd sometimes play basketball with us — and he was pretty good. I always felt we got along well. I only learned while writing this that he had Oklahoma roots. Not quite the Midwest, but close enough. Somehow he'd made the leap from Oklahoma to MIT, and eventually to CMU. Go figure.

His class drilled us in the mathematics of controlling robot arms. We had to use linear equations to calculate kinematics in six degrees of freedom — the motion of each joint so the "end effector," the robot's hand, would land in the right place. At first, it sounded impossible. Once I cracked the math (linear algebra), though, I realized it was surprisingly straightforward. Writing that code and watching a simulated arm move exactly where I told it to was a small but crucial victory. It was one of the first times I thought: *I can do this.*

But the class wasn't just about math or code. The programming was different from anything I'd done before because it wasn't abstract; it was written specifically to control a real robot arm. That meant connecting the arm to the computer, dealing with the quirks of joints and actuators, and troubleshooting when things didn't move quite the way the equations said they should. Beyond the academic lessons, I was getting comfortable around hardware and systems — the messy, real-world side of robotics. I didn't end up pursuing manipulation in my own work, but this class was another immersion in what made robotics unique: it wasn't just theory; it was engineering systems that had to actually work.

PERCEPTION, WITH STEVE SHAFER

Steve Shafer was unlike any other professor I had at CMU. He had a classic 1950s professor vibe — light blue or white short-sleeve button-down shirt, blue belt and slacks, sensible glasses, and nice loafers. He was neatly put together, even though he was a little on the husky side, but when he got excited, his energy broke through. His voice would rise in pitch, and his hands would wave around, and you could tell he genuinely loved the subject. He wasn't the kind of professor who tried to be cool; he was the kind who cared deeply, and that mattered.

Shafer was the most rigorous academician I encountered. He loved math, proofs, and structure in every sense. His perception class forced us to wrestle with how robots actually see the world. We learned the fundamentals of computer vision: edge detection, filtering, stereopsis, motion estimation. For me, it felt like magic — turning raw pixels into information a machine could reason about. The assignments weren't just theory; we wrote code that processed images and produced real results. Watching a program I wrote detect edges in a noisy photo felt like the first time I created something that *understood* anything.

One project in particular gave me a true "wow" moment. My task was to write a program that could take images of U.S. currency, geometrically normalize them by rotating and translating them into a standard orientation, and then identify their denomination. It worked. When Steve complimented me on the project — and he was not one to hand out praise lightly — I felt an enormous sense of pride. That recognition mattered. It was proof that I could hang with the best — even if they were mostly undergrads.

Although Shafer's class wasn't about autonomous vehicles directly, it hooked me. I saw how central perception was — and would continue to be — in advancing outdoor autonomous systems. It was hard, but it convinced me that perception algorithms and processing weren't just peripheral parts of robotics; they were *the* frontier. I never fully became a "perception guy," but I came away convinced that vision and perception would define much of the future of robotics.

MOBILE ROBOTICS, WITH MARTIAL HEBERT

Martial Hebert may be the smartest person I've ever met. He later became director of the Robotics Institute and eventually Dean of CMU's School of Computer Science. He had that unmistakable

French coolness about him — always calm, always composed — and he spoke with an accent he could almost seem to turn up or down depending on the room. Every so often he'd struggle with an English word, repeat it, and then just smile. He was also, to my surprise, a chain-smoker. I couldn't believe it at first. Here was one of the world's leading roboticists, ducking outside for a cigarette between lectures.

His Mobile Robotics class was both inspiring and humbling. Martial was a master at using graphics and figures to make complicated concepts simple. He wasn't afraid to get his hands dirty either — for all his brilliance, he cared just as much about making things actually work. The most important concept I took away from his course was the Kalman filter, a mathematical way to combine noisy data from different sensors to estimate position more accurately. In plain terms: a Kalman filter takes many imperfect clues about where you are — GPS, speed, steering angle — and combines them into one best guess. It was the first time I saw how math could make sense out of chaos, and I remember thinking: this is what real engineering feels like. When it clicked, it was exhilarating.

But I didn't grasp everything. One day, Martial was explaining Green's theorem to Chuck and me on a whiteboard in his office. Chuck and I just stood there nodding, pretending we understood, when we had no idea what he was talking about. We laughed about it later.

Still, the lessons stuck. Kalman filtering turned out to be one of the most useful things I ever learned, because it bridged theory and prac-tice. It was exactly what you needed for self-driving cars: taking imperfect signals and wringing something reliable out of them. Martial's later work in using laser range images to detect obstacles in the environment was a natural extension of that same mindset — cutting through the noise to find clarity.

Between Mason, Shafer, and Hebert, I got the three pillars: manipula-tion, perception, and motion. Each taught in his own style — Mason with his plaid-shirt pragmatism, Shafer with his rigorous proofs and excitement, and Hebert with his French clarity and cool — they gave me the intellectual base I'd need.

We did take 'AI' classes, but back then the field was in one of its winters. Nothing remarkable had happened in a decade or so, and I don't remember any clear path forward. Neural networks were beginning to show some promise, but they were still more curiosity than revolution. That may surprise readers now, because as I write this, ChatGPT, Grok, and other systems are in the headlines almost daily with new breakthroughs and features. The contrast between the AI of my student years and today's AI could not be starker.

But classes were only part of the equation. The bigger lessons came from the people I worked with most closely, starting with Chuck Thorpe and Dean Pomerleau — the mentors who would shape not just my research, but my path.

CHUCK THORPE

Chuck Thorpe wasn't just my advisor — he was a mentor in the deepest sense. His parents had been missionaries, and that global upbringing gave him a wide-angle view of the world. Later in his career, he would help establish CMU's campus in Qatar, but even in the early days you could see that he thought expansively, never just about Pittsburgh or CMU, but about how robotics fit into the world.

Chuck and I weren't far apart in age — maybe eight or nine years — and we had a lot in common. We both came from schools few outside our circles had heard of. We both got married young. And, of course, we both loved mobile robots. He also loved basketball, and for an "old guy," he was pretty good. We played together at the rec center more than once. I never admitted he was as good as me— he wasn't — but the truth is, he could hold his own.

As an advisor, Chuck had a knack for balancing support and direction. Each year, the faculty met to review students, and we received formal letters summarizing our progress, along with in-person meetings. I no longer have all those letters, but their language stayed with me. Chuck always put real thought into mine. They were supportive, but also clear about what I needed to do next. One year he noted that I was already producing results of publishable quality. Another described me as having the maturity of a junior faculty member.

At the same time, they never glossed over my rough spots — more than once they mentioned my impatience or that I had a strong personality that sometimes needed smoothing. Fair enough. The humor wasn't lost on me when one comment observed that I had a competitive streak that showed up even in intramurals. But through it all, the message was consistent: I was making real contributions, and Chuck trusted me to keep growing.

What strikes me now is how well Chuck was able to read me, sometimes even better than I could read myself. Left to my own devices, I would have told you that I was pushing hard, doing good work, and probably ahead of the curve. The reviews added a mirror I couldn't hold up on my own. They showed me how others saw me — talented, yes, but also impatient; driven, but sometimes too sharp in my approach.

He went so far as to note that I was moving fast — sometimes too fast for my own good — but that I was the kind of student you give room to because I thrived when challenged. That line stuck with me, because it was both a compliment and a caution. And that's the hard part of growth: you can't just evaluate yourself. You have to be willing to listen to how others see you, even when it stings, and then decide what to do with it. Chuck had a gift for putting that feedback in a way I could absorb. He never discouraged me, but he never let me think I was flawless either.

One of his key lessons was about patience. He recognized that I hadn't come from an academic powerhouse, but he also saw drive and potential. He gave me the time and encouragement to grow into the work — and to grow into myself. Those letters weren't just evaluations; they were a running commentary on my evolution from an eager but raw graduate student into someone capable of leading a major project. They mattered because I knew they came from someone who had taken the time to understand me.

And every so often, there would be a line or comment that made me laugh. It felt like Chuck slipping in a wink between the serious feedback — a note about being too competitive in intramurals, or needing to temper a strong personality. He wasn't wrong. But those small asides made the evaluations feel human. They reminded me that even in the middle of serious research, there was room for humor — and that intensity, in the right context, has its place.

I've often thought of Chuck as the father of modern mobile robots in the United States. His thesis advisor was Hans Moravec — the "grandfather" of mobile robotics — and there's a direct line from Hans through Chuck to so many of us who built what became the modern field of autonomous vehicles. Chuck led the Navlab project for nearly a decade, and so much of the technology that spun out of it shaped the foundation of today's self-driving cars. He was also the patriarch of the Navlab family. He hosted parties at his house where everyone connected to the project, even on the periphery, was welcome. That's where I first met Randy Pausch, who years later would give his famous *Last Lecture* titled *"Really Achieving Your Childhood Dreams."*

In my own view, Chuck never got the institutional recognition he deserved. When he returned from setting up the Qatar campus, I always felt he was qualified to be Dean of the School of Computer Science, maybe even Provost of the University. For reasons I'll never know, it didn't work out that way. But he landed on his feet, building a good life at Clarkson University in upstate New York, as Provost, Dean, and professor. We still trade notes every few months, laughing about the old days — and how much has changed, and how much hasn't, since the time when we were all together.

DEAN POMERLEAU

Dean Pomerleau was only a few years older than me, but by the time I arrived at CMU he had already built a reputation as one of the brightest minds in neural network-based machine learning for computer vision applications. He carried himself with quiet confidence — calm, steady, and assured.

Dean taught me a lot, not through formal lectures but by example. His system, **ALVINN**, was the precursor to much of our later work. It wasn't just a single algorithm but a full system spread across multiple computers, with separate processes handling vision, control, and display. Seeing ALVINN in action was an eye-opener. It showed me what true systems engineering looked like — how you take pieces that work individually and weave them together into something greater. That lesson changed how I thought about building technology.

Dean and I also became friends. On Friday nights, Barb and I, along with Dean and his fiancée, Terry, had a ritual. We'd go out for dessert at Gullifty's, a quirky spot in Squirrel Hill famous for its enormous pecan balls and strawberry pie, and then head back to catch an episode of The X-Files. For two young couples, it was a luxury.

He also had a crochet wall-hanging in his apartment that he'd made as a kid. We teased him endlessly about it, but it was one of those quirks that made him human in a world full of big brains and serious research. Over time, Dean and his wife Terry became some of our closest friends in Pittsburgh. We were even in their wedding in New England. Meeting his parents, I could see where his blend of toughness and kindness came from. He could be demanding, but he didn't have the sharp edge of the typical East Coast attitude, despite being from New Hampshire. He had drive, but also balance.

Dean went on to do remarkable things beyond ALVINN. While still at CMU, he developed one of the first neural-network-based eye-tracking systems. Later, he and I co-founded two businesses together

— stories I'll tell completely in a later book. And if that didn't already show his range, he even managed to train goldfish to swim through hoops for rewards. That was Dean in a nutshell: brilliant, hard-working, a little quirky, and always willing to push the boundaries of what seemed possible. For me, he was not just a colleague but a role model — close to my own age, yet reinforcing what I already knew about what it really took to succeed. Seeing someone my age embody those lessons made the path ahead feel real, and within reach.

REFLECTIONS ON CHUCK AND DEAN

I now realize how lucky I was to have Chuck and Dean side by side in those early years. Chuck gave me structure, patience, and the sense that I could grow into the work if I kept showing up. He carried the lineage of modern mobile robotics and set the tone for Navlab, not just as a project but as a family. Dean, meanwhile, showed me what it meant to think in systems — to stitch together computer vision, vehicle control, and a maze of hardware into something that actually worked. And he did it with a balance and humanity that made him as much a friend as a colleague.

Together, they formed a kind of dual mentorship. One rooted me in discipline and accountability; the other opened my eyes to creativity and systems thinking. Between the two, I began to see how I might carve out my own place in robotics — not just following in their footsteps, but starting to imagine paths of my own.

What stood out most about Chuck and Dean wasn't just their brilliance or mentorship in good times, but the way they carried themselves when things got hard. Chuck had an even-keeled way of delivering bad news. It was matter-of-fact, never dramatic, and always grounded in solutions. In the wider world, when rivals or skeptics tried to cut down our work, he projected calm, steady confidence and kept CMU looking strong. His composure under external pressure gave the rest of us room to keep building.

As our work together evolved, Dean became a defender of our vehicle-centric approach. He didn't look for fights, but he wasn't afraid to stand his ground when our philosophy was questioned. If Chuck set the tone with quiet steadiness, Dean added conviction — reminding people that faster, leaner, vehicle-first autonomy wasn't just viable — it was the better bet.

Their styles also showed up in the smaller, less visible battles of thesis work. When I was veering down the wrong path, Chuck, older and more patient, had a way of asking probing questions that led me to see the flaw for myself. Dean, by contrast, was more direct — quick to say, "That's not going to work, try this instead." At times I pushed back, but in hindsight, both approaches were invaluable. Chuck taught me to slow down and reflect; Dean taught me to adjust quickly and keep moving.

Together, they modeled two complementary ways of leading through challenge: Chuck with calm reassurance and gentle course correction, Dean with principled defense and clear direction. Watching them in those moments didn't just shape me as a researcher — it gave me a first, lasting glimpse of what real academic leadership looked like. And years later, whether mentoring the players I coached, building companies, or raising kids, I often found myself drawing on the lessons I first learned from them.

Those lessons shaped how I thought about research, but most of what I learned in those early years came from the day-to-day grind itself.

FINDING THE RHYTHM

A normal day during my coursework phase of grad school had a rhythm. I'd wake up around 8:30 or 9:00, hop on my bike, and coast downhill from our place on Wilkins Avenue to campus — usually with nothing more than a cold Coke for breakfast. By 10:00 I'd be in Wean Hall for class. By noon, lectures were done, and I'd grab lunch from the Chinese vendor in the hallway. Then it was hours in the lab, grinding through projects or homework until 5:00 or 6:00.

When I transitioned into research, the pattern shifted. Most days ran 10:00 to 7:00 or 8:00, with a break for lunch or sometimes basketball with Chuck and Matt. Nearly all that time was spent writing software. I discovered what it meant to be "in the zone" — those stretches where code flowed, pages of it, and things mostly worked. Four out of five days I found that groove, and for all the stress, there was real satisfaction in those bursts of focus where everything clicked.

And somewhere in the middle of all those classes, late nights, and half-understood papers, something unexpected happened: I started to see the path, even though I couldn't see the end of it. I began to understand just enough to realize how much I didn't know — and that was its own kind of terror. I was balancing between the fear that I wasn't capable and the suspicion that, if I kept going, I might be. It was the phase where the only real way forward was to jump into work I wasn't yet qualified to do, trusting that I'd grow fast enough to survive it.

LIFE AT HOME

Life outside Wean Hall had its own challenges. Around this time, Barb decided to go back to school too, pursuing her master's degree in Reading at the University of Pittsburgh. While I biked downhill to CMU, she walked the mile and a half to Pitt. Her fellowship meant she also had both coursework and research, often spending long hours in the library.

Barb's Perspective — On Going Back to Grad School

> *I decided to return to grad school while working in a preschool program at East Hills Elementary. Jobs were scarce, and I knew a higher degree would help. I enjoyed working with the kids, but I didn't want to stay in preschool forever. Pitt's Reading Specialist*

master's program offered both coursework and a stipend. Part of the time I was in class, and part of the time I carpooled to Highlands Elementary with other interns to work as a reading specialist. It was busy, but it felt like the right step forward.

We usually met up around 7:00 for dinner. I can't recall exactly what we cooked — though I'm sure Barb handled most of it. Two twenty-somethings, both in grad school, both juggling demanding programs and a new marriage — that was our reality. By then, the idea of kids was on the horizon. I was eager, but Barb, always the one with more foresight, said we needed to wait until life was a little more stable. She was right. Things were already crazy enough.

REFLECTIONS

The combination of classes and mentors gave me something price-less: confidence. Each project, each paper, each late-night debug session added another layer of competence. The coursework forced me to master the fundamentals of robotics. Chuck's mentorship gave me patience, direction, and a sense of belonging. Dean's systems engineering showed me how to think about the bigger picture and keep an eye on how all the pieces fit together — all while Barb and I were learning to balance the grind at school with the small routines that made our life at home work.

By the end of that year, I wasn't an expert, or even close, but I'd learned something more important: finishing didn't require already knowing everything. It required the willingness to leap before you felt ready. Every mentor who pushed me, every class that over-whelmed me, every late-night struggle made that jump a little less impossible. I didn't yet know whether I could become the kind of researcher CMU demanded — but I knew I was going to try. Some-times that's the only certainty you get before you take the plunge.

It was still daunting. There were days I felt lost or out of my depth. But I was beginning to accumulate small wins — a kinematics program that actually worked, a perception project that earned praise, a system-level insight from Dean. Those little victories added up. They set the stage for the bigger projects that would follow, and eventually for my thesis.

Those years were about learning how to build. It wasn't sophisticated. It wasn't headline-making. But it was essential. I learned the basics of robotics, I had mentors who believed in me, and I began to see myself not as a small-town kid out of my league but as a contributor to a much larger mission.

In the end, the combination of coursework and mentorship turned me from a student fumbling with compilers into a researcher ready for larger challenges. It gave me a system to fall back on: when lost, ask questions; when challenged, persist; when given opportunities, say yes. Every future rests on what comes before it. This chapter was mine — the base for everything to come.

FIRST STEPS INTO RESEARCH

LEARNING BY DOING

"Research is what I'm doing when I don't know what I'm doing."

— WERNHER VON BRAUN

---

note for the reader: This chapter leans more technical than the earlier ones. After all, this is the story of a Robotics Ph.D. I've included just enough "geek talk" to give a flavor of what the work was like, but without burying you in equations. For anyone who wants the details, full PDFs can be found online containing the algorithms, architectures, and experiments.

FINDING MY RESEARCH FOOTING

By the time I started real research, I finally had a little stability under my feet. Classes were going well, Chuck and Dean had become my anchors, and Barb and I had settled into a steady rhythm at home. For the first time, I felt a sense of momentum — enough confidence to try something new and a little daring. Classes had given me the foundation, but research was where I began to find my footing. That's true of any Ph.D. program, and especially at CMU. You don't just show up and launch into a dissertation; it takes time to learn the tools, absorb the culture, and build the confidence to tackle open problems. Like fine wine or cheese, a researcher needs a little aging before being ready. Starting research was liberating in that way: no fixed pathway, no instructions, just an open lane and enough rope to fail. It was exciting, and it was scary, and it was the first time I felt like I might be able to contribute something of my own — when the work shifted from solving assigned problems to trying to create something new.

MANIAC

My first real project was a system called MANIAC — Multiple ALVINN Networks In Autonomous Control.

During that period, ALVINN was the crown jewel of Carnegie Mellon's self-driving research. It was a feed-forward neural network trained by backpropagation to steer a car using a forward-facing camera. You'd collect thousands of images while a human drove, then train the network to map those pixel patterns to steering directions. ALVINN could drive extremely well on a particular type of road once trained, but it struggled to generalize to different road types — a network trained on highways performed poorly on local roads, and vice versa.

That's where MANIAC came in. The idea was to leverage the knowledge embedded in one network to help bootstrap another. I took the first-layer weights from multiple trained ALVINN networks — the filters that had already learned to detect image features like lane markings and road edges — and used them to seed a new network that would be trained to drive using images from multiple road types. Instead of starting from random initialization, the MANIAC network inherited a set of tuned feature detectors.

I didn't realize that reusing weights from one network to jumpstart another was an early example of what the world would later call *transfer learning* — a foundational idea behind modern deep learning. Back then, it was uncharted territory. The advantage was obvious: the new network could train faster and with fewer examples, since it didn't have to re-learn the basics of what "roads" looked like.

To make this work, I had to modify the original backpropagation library, called BP. For clarity, "backpropagation" is the algorithm that trains neural networks by sending error signals backward through the network layers, calculating how much each weight contributed to the error, and then adjusting the weights with optimization methods like gradient descent to improve accuracy.

From an implementation standpoint, ALVINN was built on BP, and the library was designed to operate on a single network at a time. I extended it to handle multiple networks in parallel, each with its own weights, training routines, and outputs. I jokingly labeled the upgrade "Hyper-Extended BP." The code wasn't pretty, but it did the job.

The results were encouraging. Using transferred weights, a new MANIAC network converged in roughly half the training time compared to one initialized randomly. More importantly, the MANIAC networks demonstrated better generalization: a system seeded from a highway-trained ALVINN was able to adapt to suburban roads with fewer failures. In controlled tests, MANIAC reduced lane-departure errors by about 20–25% compared to baseline ALVINN when moving between road domains.

What began as a simple time-saving trick turned out to be among the first published demonstrations of transfer learning — carrying what one system had learned into the next.

Publishing and presenting MANIAC was another milestone. Writing the paper was grueling — I had never written for an academic audience at that level before. The funny thing was, I wasn't particularly worried about the technical content; I knew the work was solid. What really worried me was whether I could present it effectively — whether I could explain it clearly and handle all the grammar and mechanical details that people like Chuck and Dean would catch when they proofread it. Most of the others weren't any better at those things than I was, especially the foreign students, but I still cared what my immediate peers would think. It was just another place where, at first, you feel like a bit of an outsider.

And when I stood up to present, I knew I couldn't just answer questions about MANIAC. I also had to be ready for anything about ALVINN and the Navlab project. I was representing not just myself, but Carnegie Mellon. That pressure forced me to rise to the occasion

The details appeared in a paper that Dean, Chuck, and I published at *Intelligent Autonomous Systems-3* in 1993. Reading it now, I see it as my first real contribution to the literature — the first glimmer of transfer learning in my own work, even if I was just a grad student trying to squeeze a little more life out of ALVINN.

MASPAR AND THE SUPERCOMPUTING FELLOWSHIP

Around 1992, I applied for a supercomputing fellowship. It wasn't glamorous — just a small stipend — but it came with an unusual requirement: for two summers I had to go off-site and do independent research. I chose to spend both of those summers in Denver, working at Martin Marietta. They were leading a related program that CMU was part of, so it gave me the chance to advance my own research while contributing to theirs.

Those summers gave me a glimpse into a more structured engineering world than what I experienced at CMU, but they also came with a cost. Barb was working back in Pittsburgh, and my being gone for months at a time left her holding down the fort on her own. It was one of those sacrifices we had to make — trading time together for the opportunities that would help launch my career.

However, the fellowship also came with a perk: access to exotic hardware. That meant time on a MasPar, a massively parallel computer with 16,384 processors connected in a 128x128 grid. Only about seventy MasPars were ever built, and in 1992 fewer than ten were in use worldwide. As a result, I became one of perhaps twenty people on Earth with real hands-on experience programming one. At CMU, I quickly became the MasPar guy. It was a small thing — there weren't a lot of people who cared about the MasPar — but at least in my little corner of the world, I was the guy. I was the expert. I knew more about that machine than anyone else and that was a cool feeling.

As a side note, I also had the good fortune of using Purdue's MasPar to test my algorithms, which was the largest in the world. Professor Edward Delp generously gave me access, for no reason other than kindness. He was terrific and made the experience possible. And while I'll always be a die-hard Indiana fan, it was proof that on the academic side, we could put rivalries aside — even if I still wanted to beat Purdue at everything else.

For the fellowship, the project I chose to explore was to run road-following algorithms on the MasPar. Unlike ALVINN, which was a trained neural network, this was more of a classic computer vision–driven approach. The system started by clustering pixels in RGB color space. Through a competitive learning process, each pixel was assigned to the closest cluster center, automatically grouping the image into meaningful regions — asphalt, lane markings, grass, sky, etc. Instead of thousands of raw pixels, the system now had a handful of labeled classes to work with, something that mapped beautifully onto the MasPar's grid of processors.

Once the pixels were clustered, the system used a Hough transform to detect the road. Each processor voted in "parameter space" for line segments, allowing the road centerline, lane boundaries, and road edges to emerge even in noisy data. The parallelism of the MasPar was ideal here: each processor handled a subset of pixels, contributed votes, and the collective results revealed the most likely road geometry. From there, the system estimated a drivable corridor and derived a steering command — translating clustered pixels into real control for a vehicle.

What was striking was how the hardware and algorithm complemented one another. Clustering reduced noise and dimensionality. Competitive learning adapted to real-world imagery. The Hough transform took advantage of parallelism to lock onto lane geometry robustly. Together, they formed a pipeline that could take pixels straight to steering directions in near real time.

Because we wanted to test the system on Navlab 1 — and remember this was CMU, where simulations never really counted — we had to somehow get the MasPar, itself, a refrigerator-sized machine worth hundreds of thousands of dollars, into the back of the van. I'll never forget the moment when a couple of staff members and I were wrestling that beast up the ramp, all of us silently praying we didn't drop it and end our careers in one thud. But, it was worth the effort. Sitting in the vehicle, watching road-following algorithms I had written for one of the most powerful supercomputers of its time steer, was exhilarating. Anytime you saw your software controlling a two-ton van, it was a rush.

Lab and on-road results were promising. In benchmarks, the MasPar implementations of clustering and Hough transforms were 5–10x faster than conventional workstations, fast enough to make road-following feasible in real time. But the system was also fragile. Communication bottlenecks between processors limited scalability, and the overhead of programming thousands of nodes exposed weaknesses you didn't see on paper.

That brittleness taught me humility. The system was powerful in theory, but fragile in practice. It drove home an important lesson: you can build something brilliant in the lab, but the real world will break it in ways you don't expect. That idea — robustness along with cleverness — became one of my guiding principles. And even though only a handful of MasPars were ever built, the lessons I learned about parallelism and robustness proved invaluable as I built more complex systems. Parallelism, at its heart, was just systems engineering at scale — breaking a big, messy problem into smaller pieces, managing communication between them, and making sure the whole machine worked as one. The same rules applied whether you were wiring processors together or bolting sensors onto a car.

A SIDE PROJECT: ARTIFICIAL IMAGES

Not every project I worked on was about cars or road-following. One small but memorable detour was a collaboration with a younger grad student, Shumeet Baluja, who was another of Dean's graduate students. The project explored whether neural networks could learn user preferences for computer-generated art and evolve images automatically, instead of relying on human guidance.

It had nothing to do with my main interest in mobile robotics, but I was pulled in because Shumeet wanted to use the MasPar and I volunteered to help. The paper we wrote together, *Towards Automated Artificial Evolution for Computer-generated Images*, ended up in the journal Connection Science. It was about genetic algorithms, neural nets, and aesthetics — far afield from steering a van. But that was CMU in those days: if someone needed help, you pitched in. And in the process, you often learned something unexpected.

RISING INTO THE WORK

By my third year, the lab culture had shifted. Some of the older grad students had finished and moved on, and suddenly I was no longer the new kid. I was already building a small sense of confidence — and maybe even a little expertise — in a few narrow areas. I was by no means an expert in robotics, but in certain corners of the work I felt like I knew as much as anybody. Those little victories became a springboard, pushing me to take on bigger tasks and handle larger responsibilities. Doing my own research gave me a sense of pride, even if the projects were still small compared to the big Navlab demo milestones. My confidence had grown since those first months when I barely knew how to compile code, and I was finally growing into how big an opportunity CMU really was. Here, the sky truly was the limit, and for the first time, I could feel myself rising to meet it.

At the same time, I leaned heavily on the support network around me. If I hit a wall — debugging code, choosing algorithms, framing a research question — I knew I could ask for help. The safety net was there, but now I was expected to climb higher myself. That balance — independence with support — was one of the things that made CMU such a powerful place to grow. I was finding my footing, but I wasn't doing it alone.

As I wrote my first papers and gave presentations to the larger community, I started to feel something new: a sense that the work mattered. I thought MANIAC and the MasPar experiments were good — maybe even novel — but in research you're always too close to see the impact. What brought it home wasn't my own excitement, but the way others reacted. When I stood up at a conference with Carnegie Mellon University on my slides, people paid attention. That meant what I presented had to be competent and defensible. Later, when I started to see other researchers cite my work, it was a strange and gratifying moment — not pride, exactly, but a quieter realization that I was becoming part of the broader conversation. My work, however small in the grand scheme, had become something others could build on. And that recognition — that I was finally contributing — mattered more than I expected.

LIFE AS A RESEARCHER

Once I had my own projects underway, my rhythm changed. I met with Chuck once a week to review progress, but the responsibility was mine. I had to shape research plans, study prior work, build systems, and make sense of what I found. Most days blurred together — hours of coding, chasing down bugs, refining experiments, and trying to turn the mess into something coherent. Progress rarely came in a straight line, but when it did, it felt like winning a game in overtime.

And writing — that was the hardest part. I never liked it, and I never thought of myself as a strong writer. But it didn't really matter what I built if I couldn't get it out there and let others know about it. As a grad student, that was the currency of progress. Publishing wasn't just about padding a CV; it was how you made sure the world knew what CMU, VASC, and the Navlab group were doing. And to do that effectively meant mastering something I felt weak at — something I'd barely done before. It was another skill I had to grow into, slowly, through repetition and frustration and time. I wasn't planning to be a professor, but I still had to do my share to carry the CMU banner.

Robotics also meant hardware. People rarely understand how dependent real robotics is on the physical world. You can have the greatest software in the world, but it only matters if it works on the actual machine. And with self-driving cars, those machines were huge, temperamental vehicles that demanded space, maintenance, and constant attention. None of them were designed to be robots — they were designed to be cars, and we bolted all the technology on top. That alone created endless friction. You couldn't just take them out whenever you wanted, and it felt like it was never easy. Coordinating access to Navlab 1 was a logistical puzzle: making sure it wasn't already booked, confirming the mechanical systems were ready, and hoping the weather wouldn't shut us down. Some days, rain alone was enough to keep us grounded. Other days it was something we hadn't planned for — a sensor glitch, a loose connector, or something as mundane as needing gas. Research wasn't just math and code; it was patience, preparation, and a whole lot of waiting on hardware to cooperate.

At the same time, Barb was deep into her own graduate program at Pitt, completing a master's in Reading Education. Her world was completely different from mine. While I wrestled with algorithms, vehicles, and hardware that refused to cooperate, she was studying how second-grade students understood storybook readings. The title of her thesis said it all: "The Effect of Group Size on Second Grade Students' Comprehension of Storybook Readings." And the whole structure of her program looked surprisingly similar to mine. She had a research contract — essentially her version of a thesis proposal — and a full research plan. But unlike me, she had to get written consent from parents just to run basic educational tests with young kids, while we were out driving a robot car on public roads and telling no one. The contrast always struck me as a little paradoxical.

Her work required empathy, patience, and a feel for human nuance — strengths she had in abundance and things I couldn't have done. Mine required debugging, engineering, and an ability to stay calm when two tons of steel decided not to behave. There was almost nothing in common between our topics, and yet the structure, the grind, the deadlines, and the pressure felt identical. Two young adults, newly married, both in grad school, both learning how to push through uncertainty in completely different worlds.

When I looked back at her thesis years later for this book, one thing made me laugh: in the dedication, she thanked me for helping with the computer editing. That was realistically the only part I could contribute — she did the rest, and she did it well. She finished strong, earned her master's, and in her own way was learning the same lesson I was: that deep work takes time, patience, and faith that the grind will eventually pay off. Our research topics mirrored who we were — different strengths, different temperaments — but the path through grad school shaped us together.

TAKING RESEARCH ABROAD

One of the first times I had to present our research was at a conference in Japan. For a kid from southern Indiana, even getting on the plane for the trans-pacific flight felt surreal. The flight was long, and when I landed in Tokyo, I was disoriented and exhausted. I checked into my small hotel room and turned on the TV — and somehow, against all odds, found a broadcast of Monday Night Football. This was years before streaming or YouTube. Sitting there in Tokyo, watching an NFL game, gave me just enough of a tether to home to calm my nerves.

The next morning, still jet-lagged and feeling off, I got myself together, put on a coat and tie, and headed to the conference center. Japanese breakfast wasn't what I was expecting — my stomach wasn't exactly ready for grilled fish and miso soup at 7 a.m. — but I managed to get a little down before walking over. I sat in the front row with the other presenters, waiting for my turn, trying to settle my nerves.

The room was filled mostly with researchers from across Asia, but you could feel the pressure spike whenever someone from CMU or another top institution stepped up to speak. The presentations themselves were manageable — you could rehearse those. The real test came with the questions. You never knew what anyone would ask, and you couldn't get away with stumbling around or making something up — the people in that room would spot it immediately, whether they called you on it or not. You had to stand there, think quickly, and deliver something coherent.

When my name was called, I walked up and gave the talk, and honestly, I was proud of myself. I've always had a knack for calming my nerves and presenting with confidence once I start speaking. No matter how nervous I was — and believe me, I was nervous — I could project steady control. I think some of that came from basketball, playing in front of 15,000 people and learning not to flinch. And I've never been afraid to admit mistakes or say when I don't know some-thing, which I think helps people trust you, especially during ques-tions. At the same time, if someone pushed back with something that wasn't accurate — about my results, my approach, or even the future direction of the work — I wasn't afraid to stand my ground. I learned how to shut down a bad line of questioning without escalating it, give a clear answer, and move forward. That combination — confidence, honesty, and knowing when to hold firm — made me realize I could do this — and do it successfully.

Most of the audience was kind, but I saw a few presenters get grilled hard, their arguments pulled apart in real time. It was uncomfortable to watch, but it was also a reminder that this was the major leagues. If you were going to stand up there with CMU on your slides, you had to earn it. That morning, I felt like I had.

BUILDING A PH.D. PROPOSAL

When I got back from Japan, the rush of presenting my work faded almost overnight. In its place came something more grounded, almost heavy: it was time to chart the next step. At CMU, that meant writing a real Ph.D. proposal — not a formality, but a gauntlet. You had to stand up in front of your committee and convince them you had a plan worth two, three or four years of your life, a question worth answering, and a prayer of finishing. It was equal parts roadmap and rite of passage.

For me, this was the first time I had to articulate a vision instead of just a cool demo. MANIAC and the MasPar runs had been proofs-of-concept; the proposal was essentially a contract. It committed me to a multi-year arc of experiments that would become my thesis disserta-tion. It was daunting, but clarifying. I wasn't the untested grad student in the corner anymore; I was now responsible for delivering something original, defensible, and, if I was lucky, lasting.

By the time the committee approved it, something shifted inside me. I wasn't just tinkering anymore. Their signatures meant they believed — however tentatively — that I was ready to do real research, largely on my own. In just a few years, I'd gone from barely knowing how to code to being trusted to help push the robotics world forward at one of the premier academic institutions on the planet. That realization felt equal parts scary, strange, and exhilarating.

Here's what I convinced them to let me build. The proposal was titled *Virtual Active Vision Tools for Autonomous Road Navigation.* ALVINN could already follow roads, but it fell apart on lane changes, intersections, exit ramps, and obstacles. My central idea was a framework of "virtual sensors" — software that could algorithmically shift the camera's viewpoint, feed those transformed camera images into ALVINN-style networks, and give the overall driving system a way to reason about harder tactical situations.

Technically, it broke down into three threads:

1. **Algorithms** – developing ALVINN-based modules that could handle tasks beyond simple lane-keeping. For example, an intersection detector that could shift the field of view and classify road geometry in real time.
2. **Integration** – defining how these modules would plug into a broader driving system. I had to show not only that each piece could work but that together they could form a coherent architecture for autonomous navigation.
3. **Evaluation** – proposing how to test and measure performance. That meant specifying datasets, test sites (like Shenley Park and later Breezewood), and metrics such as detection accuracy, response time, and lane-following error.

The committee grilled me on feasibility — data collection, hardware limits, scope. I walked out with their approval — and the quiet, electric realization that the next mountain was now mine to climb.

REFLECTIONS

Research didn't come with a roadmap. After years of working under others in the lab, suddenly the direction was mine to set — and that was both liberating and intimidating. I began doing what I'd call "innovative research," or at least as innovative as I was capable of at that stage. The freedom felt good, but it also came with the weight of knowing there was no guaranteed outcome. I had enough rope to fail, and that alone pushed me to grow. These early projects weren't about glory. There were no magazine articles or headlines. They were about grinding in the lab, writing code that didn't work until it finally did, presenting papers to skeptical audiences, and slowly transforming from a student into a scientist.

Over time, I accumulated small wins — a program that finally worked, a system design that actually held together, the slow shift from being the youngest person in the lab who needed help to one of the older ones who could give it. That transition, from dependent to contributor, was its own kind of growing-up moment and gave me a kind of confidence I hadn't had before. A* taught me how to code, how to survive in Unix, and how to fight through failure until something worked. MANIAC gave me confidence that I could contribute something novel. MasPar forced me to think differently and value robustness. Moving up in the lab showed me I could carry responsibility, not just lean on others. Seeing citations of my work gave me a first taste of real impact. Working with Shumeet reminded me that collaboration mattered, even outside my lane. And the Japan trip showed me what it meant to stand on a global stage.

Publishing my first paper during this period was another milestone. The writing was difficult, the review process was stressful, and presenting it was downright scary. But once it was done, I'd completed the full arc of what research required: have an idea, work through it, produce something that genuinely advanced the field, and defend it. That cycle would become the backbone of my eventual thesis. None of this was easy. There were plenty of mundane

moments that nudged me into territory I hadn't been in before —
contacting a vendor for the first time, reaching out to a researcher at
another university, even applying for the Supercomputing fellowship.
Each of those steps felt small on the outside but big on the inside.
Every one of them pushed me a little farther, expanded my network,
and taught me a new skill I didn't know I needed.

The work was technical, but the lessons were human: persistence,
humility, resilience. You can't be afraid to dive in over your head. You
can't be afraid to ask questions. And you can't fool yourself into
thinking clever is enough — in the end, things have to work in the
real world. The outcome of all this was simple but profound: I began
to believe I could take on bigger challenges. Whatever direction my
thesis took, I felt capable of doing the work. These years completed
the emotional and intellectual foundation that had been forming
since my first semester, and prepared me for the leap that was coming
next: putting everything together in a minivan, pointing it west, and
driving across America with no hands on the wheel.

NO HANDS
ACROSS AMERICA

THE JOURNEY THAT
CHANGED EVERYTHING

When did the idea begin? The exact moment has blurred with time, but I remember a drive back to the Denver airport from a demo at Martin Marietta with Dean and Chuck, when the conversation shifted from ALVINN's promise to its limitations. It had taken us far, but it couldn't adapt quickly enough when conditions changed. Something new had to come next, and it had to come quickly, with the Automated Highway System (AHS) demo already on the horizon.

At that point, there was no No Hands Across America — only the recognition that whatever replaced ALVINN had to be tested rigorously and at scale on real roads. Out of that need for massive testing, the idea of a cross-country run eventually took shape. Bold, yes.

Different, absolutely — but not crazy. If self-driving cars were ever going to be real, somebody had to show the world it could be done. Somebody had to drive across America, no hands.

We saw it as a technical necessity — the only way to gather the miles we needed. What I didn't see, what none of us saw, was the way this moment would echo far beyond research papers and demos. I had no idea that decades later, this drive would become a story I'd tell my kids — and eventually, my grandkids — as part of our family history.

FROM ALVINN TO RALPH

To understand why No Hands Across America worked, you need a quick tour of how ALVINN actually saw the road — the good and the bad. Don't worry, no math exam here. Just enough detail to show why the approach mattered.

For several years, ALVINN had been our flagship lane-keeping system. It could learn remarkably well to drive on a specific kind of road, and when we combined multiple trained networks through MANIAC, the system could even handle different environments reasonably well. But the MANIAC approach carried a fundamental flaw: it was like teaching a student every answer on the test instead of teaching them how to think. Unless a stretch of highway matched one ALVINN had already memorized, it could get confused. And the world contained too many variations — in surface, texture, markings, and lighting — for us to realistically cover them all. ALVINN could be retrained in just a minute or two — remarkable for its time — but that still wasn't nearly fast enough at highway speeds, where conditions could change in just a few miles, or even instantly.

That limitation led to one of the most important insights of the entire project. On that drive in Denver, Dean and Chuck began talking about how roads themselves offered inherent structure — strong parallel lines, consistent directionality, and edges that naturally guided the eye forward. Instead of forcing a network to memorize every possible scene, what if we built a system that could instantly lock onto those visual patterns as they appeared? I mostly listened, chiming in here and there, but even in the moment it was clear: this was the conversation that set the direction — the insight Dean would soon turn into RALPH.

RALPH — the Rapidly Adapting Lateral Position Handler — represented a real shift in our thinking. To appreciate that shift, you have to understand where we came from. Dean had spent his entire academic career building neural-network-based systems — training them to follow roads, recognize structure, and generalize across conditions. ALVINN, MANIAC, and our lineage of work before RALPH were all grounded in learning-based approaches.

But RALPH wasn't neural-network-based at all. It still used simple learning techniques, but the core idea flipped the script entirely: instead of depending on a network to discover the right features, RALPH encoded the universal *structure* of roads directly. It was, in many ways, the opposite philosophy — taking what we had learned from neural-network experiments and improving on them using explicit domain knowledge. That distinction mattered. It showed that progress didn't require loyalty to a method; it required loyalty to what worked.

And that was part of CMU's culture. The chase was always for the best solution, not the most fashionable one. You didn't have to stay locked into a particular paradigm. If you found a better way — whether it came from a neural network, a hand-engineered model, or something in between — you were encouraged to follow it. I can't speak for how other institutions operated, but at CMU, that freedom to explore alternatives wasn't just allowed. It was expected.

Technically, RALPH took advantage of a simple truth: roads almost always contain strong visual cues running in the direction of travel. They have to; those cues are what make human driving possible. Neural networks are powerful because they can discover patterns a person couldn't easily code. They generalize well. But that flexibility comes at a cost. Once a network locks onto a representation, it becomes difficult to change what it pays attention to. That made ALVINN brilliant on a single stretch of pavement, and brittle when the world changed.

My earlier work with MANIAC had run into the same roadblock. I wanted a system that could handle any road, in any condition, but I focused only on learning — not on manually encoding the common structure we knew every road possessed. RALPH took the opposite approach, and that's why it worked. By modeling that universal structure explicitly, it adapted almost instantly to new surfaces, markings, and lighting. It solved the exact limitation that had held ALVINN back.

Once Dean built the first version of RALPH, we knew it had to be tested — and tested fast. The Automated Highway System demo was coming, and RALPH needed to be ready. That urgency is what set the stage for No Hands Across America. Local trials weren't enough anymore. If we wanted to validate RALPH properly, we needed massive amounts of data across the widest possible variety of roads. And what better way to get that than driving coast to coast?

Someone had to show that self-driving cars could handle America — all of it. And the only way to prove it was to go out and do it.

From a vantage point more than 30 years later, what began as a data-gathering run now feels like something much bigger. No Hands Across America (NHAA) wasn't just about training RALPH — it was about showing, for the first time, that self-driving cars could leap off the test track and onto America's highways.

THE RIVALRY

There was another force pushing us forward: competition. By the mid-1990s, the federal Automated Highway System program was just coming online, and our group at CMU was part of the winning consortium. But even inside that victory, it was clear who carried the most weight. Partners for Advanced Transit and Highways (PATH) at Cal-Berkeley stood as the dominant force in the community. With major funding and deep ties to state and federal transportation agencies, their demos came off as polished showcases — tightly choreographed, strategically staged, and designed to signal leadership.

Our group at CMU was the opposite — smaller, scrappier, and wired directly into DARPA and the Department of Defense. PATH looked like the frontrunner; we were the underdog. And if they were going to play it safe, we were going to do the opposite: roll out something bold, different, and — at least in our minds — entirely reasonable. A minivan driving itself across America. In many ways, No Hands Across America became our opening salvo inside the consortium — our way of signaling that we were here, and that we intended to contribute in a very different way. That contrast in styles didn't just fuel NHAA; it set the competitive backdrop for the AHS demo that would follow.

Of course, it's entirely possible that I conjured up much of that rivalry myself. I've always believed you sometimes need an enemy to do great things, and maybe I simply cast PATH in that role. I doubt they ever saw it the same way. Maybe they did, maybe they didn't. But in my mind, and in the minds of many of us at CMU doing the day-to-day technical work, that's exactly how it felt.

We needed someone to measure ourselves against. It was the same instinct I'd carried from my days in sports — competition sharpened you, pushed you harder, and made victories that much sweeter. And it wasn't lost on me that PATH was Cal-Berkeley. I was at CMU, but I had come from Indiana State, a kid whose classmates had gone to places like Berkeley. That chip on my shoulder made the rivalry even more real. And just like in sports, the mindset never changed: the goal was always to win — and, if possible, to make winning feel inevitable.

RISK AND READINESS

Before we could make that vision real, we had to convince the university it was even safe to try. Takeo and Chuck went to the administration and trustees, making the case that putting experimental robots on highways wasn't reckless but exactly the kind of challenge CMU was built for. After all, the Robotics Institute itself had been created on that same premise — that bold ideas had to be tested in the real world, not just on paper. In that sense, the DNA of CMU ran straight into our project: a willingness to take risks, to back its people, and to prove technology outside the lab.

The trustees gave us the green light — but no safety net. We were cleared to try, and succeed or fail, it would be on us. I can only assume Takeo and Chuck soft-pedaled the risks (or maybe fibbed a little — joking, of course) to get that approval. But the truth is, it took real courage on their part — and on the part of the university leadership — to sign off on something like this. They had to trust us, trust the science, and trust that the institution's mission mattered more than playing it safe.

Looking back, I'm not sure in today's world — with all the scrutiny, the liability concerns, and the instinct to protect multi-billion-dollar endowments — that something this bold would ever get the same blessing again. The fact that it did then speaks volumes about the leadership CMU had at that moment. In later books, I'll show how this kind of institutional courage became harder to find in similar places over the following thirty years.

The vehicle that would carry us across the country came courtesy of Delco Electronics, thanks in large part to a 30-something engineer named Ashok Ramaswamy, who was Delco's AHS lead. Ashok had a dry, sharp wit that could catch you off guard, and he was genuinely funny. More than that, he liked Dean, Chuck, and me — and the feeling was mutual. I always sensed that he bristled at the way PATH operated, with its vision of embedding millions of magnets in roadways. He, like us, believed the future of self-driving cars had to be on-board intelligence, not infrastructure.

Ashok believed enough in our approach that he handed us a silver Pontiac Trans Sport, no strings attached. Navlab 5 was ours to do with as we pleased. The GPS receiver and fiber-optic gyro came from other sponsors. The laptop and camera may have been borrowed or donated as well. All told, the equipment probably cost $20,000, scraped together from overhead and favors — an absurdly small number by today's standards. (For the record, there were Navlabs 2, 3, and 4 in between 1 and 5 — each its own stepping stone — but it was Navlab 5 that really defined this era for me.)

The logistics were minimal — almost laughably so. We had an AAA TripTik booklet for the route, a truck scheduled to haul the van back to Pittsburgh at the end, and not much else. No chase car. No live tracking. No PR team. Not even hotel reservations. Just us and the open road.

It's almost comical to imagine pulling off something like this today. In the age of social media, the entire trip would have been livestreamed, tweeted, TikTok'd, and dissected in real time. Every hiccup would have been magnified, every mile scrutinized. And just as loudly, there would have been the chorus of critics: people asking why a "robot car" was allowed on public highways, raising safety concerns, worrying about liability, and demanding oversight.

Back then, it was a simpler world. Without the glare of constant coverage and public outrage from one side or the other, we were free to focus on the work itself — a couple researchers chasing a big idea, armed with little more than a van, some borrowed gear, and a stubborn belief that it could be done.

The idea of posting updates to the Internet was a last-minute thought, mainly for colleagues back at CMU. The Internet wasn't a mainstream phenomenon yet, but we decided to give it a try. Each night, after a long day on the road, we'd spend at least an hour —

sometimes more — writing journal updates, editing raw HTML in emacs, and uploading text and photos via FTP. I'm not even sure why we put that much effort into it, but 30-plus years later, I'm glad we did. We even tried to add a little humor here and there, figuring maybe someone back home was actually reading. At the time, it wasn't meant to be a global broadcast. But those short nightly entries turned into one of the earliest blogs ever created.

Technically, the trip was remarkably stable. Once we left Pittsburgh, there were no major code changes that I can remember. No debugging marathons. Maybe a few parameter file tweaks, but otherwise it was what it was — a real test of RALPH in real conditions. It wasn't about code anymore — it was about proving RALPH on real roads.

SAFETY

We weren't oblivious to the fact that what we were doing could be dangerous — and maybe even a little reckless — so we knew we needed at least a rudimentary way for the human driver to retake control. From the start, we assumed the system wouldn't work 100% of the time, so the first line of defense was always the human. The steering motor was deliberately underpowered so a driver could overpower it instantly just by grabbing the wheel.

On top of that, I wrote monitoring software for an HC11 microcontroller that constantly compared the commanded steering angle to the actual wheel angle. If they diverged too long or too much, the motor disengaged automatically. And of course, we had the big red button on the dash — press it, and RALPH was instantly out. It's funny to think about now: an 8-bit HC11 running under 2 MHz, with just a few hundred bytes of RAM, was the guardian of our lives. Today's smartphones could out-compute it by a factor of millions.

Relying on the driver as the ultimate backstop also exposed the biggest flaw in that approach: people make mistakes, and attention drifts. We knew that even then, but it was the best we had. Modern systems don't get that luxury. With human attention no longer assumed to be reliable, the autonomy itself has to be far more robust and fault-tolerant. From perception to decision-making to actuation, every layer now needs to be hardened, redundant, and designed to fail gracefully.

That reality was underscored a few years later during AHS demo testing, when one of PATH's practice runs went wrong. Vehicles traveling too closely in formation collided because the human safety drivers simply didn't have time to react when a failure occurred. It was a glimpse of the limits of their approach — and a reminder that robustness would ultimately matter most.

THE ROAD AS OUR DATASET

Tesla now gathers billions of miles of data automatically from its fleet every year. In 1995, the only way to generate data was to physically drive the miles ourselves, recording it onto DAT tape recorders strapped into the back of our test vehicles — high-tech for the time. That's what it took. And even then, nearly all of our data came from the Pittsburgh region and the same handful of roads we happened to drive on. There were no massive online libraries of road footage, no simulation environments, no crowdsourced driving logs. That scarcity was exactly why No Hands Across America had to happen at all. It was our way of breaking out of that box — taking the system coast-to-coast so we could capture something bigger. The trip probably increased our training data by an order of magnitude over earlier projects, but even then we still had millions — even billions — of times less data than modern companies use today.

The contrast between the early days and today is staggering. In the 1990s — during the No Hands Across America and AHS era — we logged only a few thousand miles in total, and each one required hands-on engineering, late nights, and a fair amount of luck. Our available sensor suite was primitive by modern standards: a low-resolution camera, an early radar or laser scanner, and the first generation of GPS. Everything we captured was stored in megabytes of data, archived to tape or slow hard disks, and processed on a single on-board workstation that could barely keep up in real time. Each vehicle cost millions and depended on hand-engineered features drawn from small, carefully curated datasets.

Three decades later, the landscape is unrecognizable. Companies like Tesla, Waymo, and Cruise have collectively driven more than ten billion real-world miles — and trained their systems on hundreds of billions more in simulation. Fleets bristle with HD cameras, LiDAR, radar, ultrasonics, and precise GPS/IMU units, feeding petabytes of data into global cloud warehouses. The compute power that once filled a van now lives in processors capable of training trillion-parameter models. What once cost millions per vehicle now runs on commodity hardware for a fraction of the price. And instead of hand-crafted features, today's systems learn continuously, retraining themselves on vast data streams flowing in from around the world.

That was the gap. In the '90s, we were data-starved pioneers, wringing meaning out of every mile we could record. Today's companies are data-rich industrialists, training at planetary scale. That's why No Hands Across America mattered so much: it gave us the only thing we couldn't simulate or shortcut back then — real miles, under real conditions, on real roads.

HOW THIS STORY WILL BE TOLD

Unlike previous chapters, which I've written from memory and reflection, this one will unfold in a slightly different way. As I mentioned earlier, Dean and I kept a nightly online journal. Each evening we typed up the day's events, recorded the autonomy statistics, and shared the human details — from construction zones to Burger King stops to odd roadside attractions.

Below, I present that original journal verbatim, in italics, exactly as it appeared in 1995, **typos and grammar errors included.** That preserves the immediacy of what we experienced — the voice of two researchers describing what was happening, often with humor, sometimes with fatigue, always with a sense of discovery. Also included are the original photos from the trip, or similar duplicates if the originals were not available. **Please excuse their quality.** It was 1995!

After those journal passages, I've added brief modern reflections and commentary — to explain technical context, the significance of what we were seeing, and the personal meaning those moments have for me now. The journal gives you the *what happened*; the commentary offers the *why it mattered*.

And so, with only a barebones plan, a few sponsors, and the conviction that somebody had to do it, we set out to see if a minivan could drive itself from Pittsburgh to San Diego.

PROLOGUE TO THE JOURNEY: PITTSBURGH TO WASHINGTON, D.C.

About a month before we headed west, we ran a full dress rehearsal to the east. We drove Navlab 5 from Pittsburgh to Washington, D.C., where RALPH handled about 96% of the 290 miles autonomously, including a stretch close to 90 miles without interruption. The trip had a political purpose: the Automated Highway System program had started, and we wanted lawmakers and agency folks to see the technology firsthand.

This was pre-9/11 America: we parked right on the street between the Rayburn and Longworth House office buildings, popped the back, and showed people what we'd built. Security never crossed my mind — what struck me was how cool it was to be standing there talking with actual U.S. Congressmen and Congresswomen. It felt like an honor, something I don't think I would necessarily feel the same way about today. In the strangest moment of the day, the driver's side window suddenly shattered while we were parked under a tree — no rock, no bullet, just a loud crack and shards of glass. Today that would spark a full-blown security panic; back then we swept it up and kept talking.

That D.C. run was the first act of No Hands Across America — a proof that our coast-to-coast story already stretched from the Atlantic side. In a few weeks, we'd write the western half.

DAY 1: JULY 23RD, 1995 – PITTSBURGH TO INDIANAPOLIS

Driving Time: 7:50am - 2:00pm

Segment Autonomy Stats:

- *Autonomous Driving Percentage: 99.2% (381.12 / 384.38 miles)*
- *Longest Autonomous Segment: 46.98 miles*
- *Avg Speed: 59.6 mph*

Total Autonomous Stats:

- *Autonomous Driving Percentage: 99.2% (381.12 / 384.38 miles)*
- *Longest Autonomous Segment: 46.98 miles*
- *Avg Speed: 59.6 mph*

We departed Pittsburgh without much fanfare on a sunny Sunday morning. The departure was somewhat anticlimactic after the hectic preparations over the last couple of weeks. The first hour was uneventful, and we crossed into West Virginia at 8:46am. Another 15 minutes passed and we crossed into Ohio at 9:00. The states are going by quickly! During this stretch we traversed our first (of many) construction zones. RALPH handled them admirably, for a one-eyed four month old :-). After our first 60 miles, we were 98% autonomous.

To help us stay alert, we brought along a number of books on tape. Today we listened to Star Trek "Generations", which was pretty mediocre. Entering into Ohio, the road began to look like it will for most of the trip -

open fields, long straightaways, lots of trucks. We made fairly good time to Columbus, considering there was a 10 mile detour onto rural state highways. RALPH autonomy stats remained high, around 98%.

We reached Columbus at 11:15am. Shortly before reaching Columbus, we got a surprise call on the cell phone from Doug Pape, a collegue at Battelle Memorial Institute. He and his family had just left Sunday school and wanted to catch of glimpse of Navlab 5 in action. We arranged a rendezvous with them on I-70 West of Columbus. They caught up to us in their mini-van, and drove alongside for a few miles. We waved, they snapped a few pictures, and then we were on our own again. But before leaving the Columbus area, a reporter from a local TV station phoned wanting to do a quick interview to supplement the video they got in Pittsburgh on Friday. It was a nice break from the monotony of the open road.

Things were going so well, we decided to push on without stopping. On the road from Columbus to Indianapolis, we started to achieve some long autonomous stretches. During the next two hours, RALPH drove without intervention for several 15+ mile segments, including one of over 45 miles. These included several sections of construction, as well as pretty heavy traffic.

We reached Indianapolis at 2pm, and headed for the Indianapolis Motor Speedway, home of the 500. We took some hurried photos at the entrance, and then headed for the visitor's center. In the visitor's center we saw what we hope will be the next generation navlab vehicles - a NASCAR stock car and the 1995 Indy 500 Corvette pace car. After driving around the track [in a tour bus :-( ], we took a few more pictures and left.

Having not eaten since 7:30am, by this time we were starving. We decided on Burger King, and by chance stumbled on what we decided should be the unofficial mascots of the "No Hands Across America" Tour, the Disney/Burger King Pocahontas figurines. We purchased the first of many kid's value meals, and were presented with a statuette of the beautiful leading lady. After a moving ceremony, we firmly duct taped her to the dash, as our guiding light for the rest of the journey.

With the days festivities complete, we headed to the Northern part of Indianapolis, so we'd be ready for tomorrow's visit with one of our primary sponsors, Delco Electronics.

REFLECTIONS ON DAY 1

The night before we set out, I expected to feel anxious — maybe even overwhelmed by the uncertainty of it all. After all, there wasn't exactly a playbook for driving across the country without touching the wheel. But instead, I felt surprisingly calm. We'd tested the system thoroughly. We had a safety driver. And deep down, I believed it would work. It was an exciting time — the kind of anticipation that comes when you know you're standing at the edge of something no one has ever done before. Maybe it's overstating it to compare us to Columbus crossing the ocean, but the feeling wasn't far off. We were about to attempt something unprecedented, and that sense of possibility stayed with me through the night.

We did get a warm sendoff outside the Robotics Institute that morning, but once the minivan merged into the interstate stream, the moment settled into something quieter, almost ordinary — which was its own kind of thrill. Construction zones provided the first real test. RALPH handled them, while our hands hovered inches from the wheel and our eyes moved constantly, no differently than any attentive human driver.

By then, the work was guided less by conscious calculation than by feel. After so many hours around autonomous vehicles, it became possible to sense when a situation was routine and when something was slightly off. The road ahead was scanned, and almost instinctively, there was an understanding of what RALPH could manage. When conditions looked clean and predictable, hands relaxed away from the wheel. When something seemed uncertain — an odd curve, degraded lane markings, a car edging too close — they returned without deliberation. Certain conditions always demanded heightened attention: driving into a low, setting sun, traffic compressing unexpectedly, a vehicle cutting in sharply.

Most of the time, though, the system remained stable. The role shifted from vigilance to observation. Trust gradually replaced tension. And on occasion, the experience even grew boring — a quiet, unexpected signal that something extraordinary was actually working.

The Burger King ritual started as a joke and morphed into a rhythm. On a trip like this, tiny traditions — like duct-taping a plastic Pocahontas to the dash — make the miles feel human.

DAY 2, PART 1: JULY 24TH, 1995 – INDIANAPOLIS TO ST. LOUIS

Driving Time: 2:30 pm - 6:45pm

Segment Autonomy Stats:

- *Autonomous Driving Percentage: 98.7% (239.05 / 242.18 miles)*
- *Longest Autonomous Segment: 47.03 miles*
- *Avg Speed: 58.05 mph*

Total Autonomous Stats:

- *Autonomous Driving Percentage: 98.98% (620.17 / 626.56 miles)*
- *Longest Autonomous Segment: 47.03 miles*
- *Avg Speed: 59.0 mph*

The day began in Kokomo, IN with a visit to Delco Electronics, one of our primary sponsors. After giving a talk to about 50 Delco engineers, we demonstrated the system autonomously steering the vehicle as well as operating in a lane departure warning capacity. The talk seemed to be well received and we received many good questions and comments.

After a quick lunch with Ashok Ramaswamy, our liason at Delco, we headed back to Indy to resume our trip. We decided to test RALPH on the way back to make sure it was set up properly for interstate driving. At this point we discovered what we have come to call "The Curse of Pocahontas." During our testing, RALPH was driving erratically, but we couldn't figure out why. We were getting desparate, but then we noticed that our guiding light was(n't) affixed to the dashboard. (We had taken her down for the demos.) Just after affixing to the dash, we noticed that the steering wheel encoder connector had fallen off. This was very strange, since minutes before lunch, the system had worked perfectly.

After reattaching the connector, RALPH was back to normal. We suspect that Pocahontas and her sidekick, the energizer bunny (who was also sealed in the wooden computer platform) had conspired to somehow disconnect the cable. Or it could have just wiggled loose. But from now on, we are taking no chances. Pocahontas will always be firmly attached to the dash with her sidekick nearby.

During the drive between Indy and St. Louis, we contacted The Tonight Show with Jay Leno. Mr. Leno had expressed interest in seeing the vehicle and we wanted to set something up. (He did a joke about Navlab 5 a couple of weeks ago.) Although we're not on the show, we are hopeful that Mr. Leno will be able to take a look at the vehicle and maybe mention the trip during his monologue.

Of special excitement to Todd was the opportunity to drive through Terre Haute, IN again. I spent my undergraduate days there and hadn't been back in a while. Although we didn't get to go through the campus, it was nice to see the scenery again. I pointed out the creosote plant to Dean, who thought it was "quite picturesque. And smelled nice too."

Things were going smoothly until we reached Effingham, IL. There was some unmarked construction (on the AAA map) and it was very frustrating. During this time we had the opportunity to interview 3 guys from Ohio who were heading to Wichita, Kansas. We were stuck next to them in traffic so we decided to show them the vehicle driving itself. They thought it was pretty cool and it passed the time for us.

The section from Effinham to the outskirts of St. Louis went well. (We drove by Pocohantas, IL.) We surpassed our previous longest autonomous segment by a hair with a 47.03 mile stretch without touching the wheel. We were averaging about 99.4% autonomy until just before the Mississippi. At this point several factors conspired to lower our final percentage to 98.7%. First, the road and road markings deteriorated substantially. Next, the sun was getting low on the horizon, causing some specular reflections. And finally, it had started to rain.

We made it to the western edge of town before we decided to take a break for dinner and some sightseeing. In the excitement of acquiring our second

Pocahontas action figure (John Smith), Dean somehow lost the entry key to Navlab 5. We searched our pockets, the restaurant, and the the parking lot, but it was nowhere to be found. Fortunately, Todd's Boy Scout background came to our rescue. He had remembered to tape a spare set of keys under the rear bumper. We retrieved them and were on our way, vowing to be more careful in the future. We headed downtown to the Gateway Arch for some quick photos. The sun had now set for the evening so we decided to fill up the tank and head for Kansas City.

REFLECTIONS ON DAY 2, PART 1

The Delco engineers were generous with their attention — engaged, curious, and realistic about the gap between a research van and a production car. When the encoder plug shook loose, the relief of a simple fix reminded us how much of this trip depended on banal details staying put. The contact with Leno's team added a Hollywood sheen, but we were still two guys improvising our way through the Midwest. Effingham's traffic jam turned into an impromptu demo for three Ohio travelers; they didn't see a gimmick, they saw a glimpse of what might be coming. And those final miles into St. Louis highlighted a truth that still lingers three decades later: low sun angles and rain remain kryptonite for vision-based systems. Just the other day, driving a rental car, the dash suddenly beeped at me as the setting sun blinded its sensors — *"Forward sensing systems unavailable."* It was the same fragility we'd seen in '95 — proving that some challenges evolve, but never fully disappear. Back then, we were fighting them with RALPH running in a minivan. In 2025, automakers are still fighting them with fleets of cars and supercomputers.

DAY 2, PART 2: JULY 24TH, 1995 – ST. LOUIS TO KANSAS CITY

Driving Time: 8:30 pm - 12:15am

Segment Autonomy Stats:

- *Autonomous Driving Percentage: 96.5% (203.44 / 211.02 miles)*
- *Longest Autonomous Segment: 25.47 miles*
- *Avg Speed: 60.38 mph*

Total Autonomous Stats:

- *Autonomous Driving Percentage: 98.33% (823.61 / 837.58 miles)*

- *Longest Autonomous Segment: 47.03 miles*
- *Avg Speed: 59.3 mph*

This segment was the first night segment of the trip. RALPH was having problems initially due to poor lane markings and improper low beam headlight alignment. The left headlight is aimed a little too far left, leaving a large dark area about halfway up the left side of the image region that RALPH uses. For the road we were driving on, this meant that the dotted white lane marking did not appear in the RALPH image until it was almost unusable. We solved the problem by driving with the hi-beams on. Suprisingly, only a couple of cars on the other side of ther interstate even noticed that they were on. It seems that we are headlight deprived.

The last part of the trip was pretty frustrating. The road was freshly paved, but the road workers went home before they got a chance to finish all/any of the line painting. Only short dots, where the normal dashed centerline would be, were present. Needless to say, this affected RALPH's performance, but we made it through OK.

During the 210 mile trip, we passed the 1000 total miles driven mark. This included both miles driven autonomously, and those driven to make visits and see sights. In any case, it's a lot of miles to drive in two days. We were quite pleased that RALPH was also able to steer the vehicle without intervention for several 20+ mile stretches, including one of over 25 miles. Although RALPH's overall performance was the lowest yet, the conditions were also the most challenging. More to follow on this segment later, but we are both pretty tired.

REFLECTIONS ON DAY 2, PART 2

Night driving in 1995 showed autonomy was possible, but only barely. It was on the edge of what felt safe or reliable. Small things that a human driver would shrug off became big problems for the system. Misaligned headlights made driving very difficult until we switched to high beams. Reflective glare made the painted lines shimmer and disappear. Worst of all were the stretches of fresh blacktop where the highway crews had laid pavement and left nothing but a dot of paint every hundred yards or so. A human could easily interpolate the lane, but for RALPH it was nearly impossible. We managed, but it underscored a truth we learned again and again: autonomy isn't just about the car or the code. The environment has to cooperate at least a little.

And on top of all that, I want to point out something people often miss when they hear the story: we drove roughly a thousand miles in just two days. We weren't truck drivers, and we weren't used to that kind of pace. The physical grind alone was exhausting, but layered on top of it was the constant, low-level stress — sometimes explicit, sometimes just humming underneath — that we were testing something potentially dangerous, definitely novel, and never tried before. Every mile carried a kind of tension that wore on us. But we pushed on. Part of it was the excitement of what we were doing; part of it was our youth, which has a way of cushioning fatigue you don't even realize you're carrying. Whatever the mix was, it kept us moving forward.

DAY 3: JULY 25TH, 1995 – KANSAS CITY TO DENVER

Driving Time: 6:30am - 5:50pm

Segment Autonomy Stats:

- *Autonomous Driving Percentage: 98.65% (635.41 / 644.13 miles)*
- *Longest Autonomous Segment: 68.98 miles*
- *Avg Speed: 66.2 mph*

Total Autonomous Stats:

- *Autonomous Driving Percentage: 98.49% (1461.72 / 1484.13 miles)*
- *Longest Autonomous Segment: 68.98 miles*
- *Avg Speed: 62.19 mph*

After getting about 4 hours sleep, we arose bright eyed and bushy tailed at our Ramada hotel outside of Kansas City. We met with a reporter from Business Week named Otis Port who was going to observe part of the trip for a story on No Hands Across America. The segment started out a little shaky, with a power glitch before we got on the highway. But after a quick reboot, we were off.

The section around Kansas City was actually the toughest daytime stretch we'd encountered thus far. Heavy rush hour traffic, coupled with construction and narrow lanes, resulted in a lower initial autonomy percentage. But we made it up on the interstate between Kansas City and Topeka, driving 94.45 out of 96.42 miles autonomously (98%). This included a 32.5 mile uninterrupted stretch.

We dropped Otis off at the airport in Topeka at about 8:30am to pick up his rental car (his flight was out of Kansas City later that afternoon). He mentioned trying to catch up to us on the highway for some quick outside photos, but once we had gone about 125 miles we gave up looking for him. But then he was there beside us. He took a couple of shots while driving next to us, and then sped ahead to so that he could get a good photo from the shoulder of us passing by him.

At about this time, we passed the sole state trooper we saw all day. In our rear view mirror we saw him pull out. We quickly performed an autonomous lane change, nervously waiting for him to pull in behind us. Much to our relief, he passed us by. Otis was not so lucky. About a mile ahead, we saw the state trooper pull up behind Otis, who was already on the shoulder and out of his car. As we approached, Otis ran by the cop, who was obstructing his view. Otis pulled out his camera and snapped our picture as we drove past waving. We crested a hill and that was the last we saw of Otis and the trooper. During the next 4 hours, we covered 289.75 miles, 99.4% of it autonomous. This stretch included the longest autonomous segment thus far, 68.98 miles.

Like an oasis in the endless fields of wheat, we saw Prairie Dog Town on the horizon.

We needed gas anyway, so we decided to check it out. We payed our $4.75 admission, and entered the most incredible menagerie of creatures to be found on I-70. The tour started off with a box full of rattle snakes. The proprietor, noticing we had cameras in hand, wanted to give us our money's worth. She opened the box and said, "If you're gonna take pictures, I'll stir 'em up a bit for you." She proceeded to poke at the rattlers with a sharp stick, while we rolled the video.

We went out back, and walked past about 50 cages full of pigeons, until we reached one of the star attractions advertised on their giant roadside sign, an 8000 lb. prairie dog. To our disappointment it was only a statue - we had been suckered by a classic bait and switch.

Figuring the whole place was a scam, we walked dejectedly back to the entrance. Passing the proprietor, we asked if the other star attraction, a 6-legged steer, was on vacation. She said, "No, he's back there in the pen." Still skeptical, we asked if she had sewn the extra legs on. She responded, "No, I wouldn't have done that bad a job." By her tone we could tell she was sincere, so we hurried back to check it out. Lo and behold, it was really true, there stood a 6-legged steer. Evidently, the extra appendages made the poor beast somewhat of an outcast, as its other pen-mates bullied it as we tried to pose for a picture. After spending nearly an hour, it was time to go. We thanked the proprietor and left, feeling we had gotten our money's worth.

Our next stop was Burger King, where we collected the next of our Pocahontas action figures, Percy the dog. We ate quickly and resumed the drive to Denver. The next 3 hours were pure excitement - 207 miles of Kansas and Colorado prairie. RALPH drove 99.2% of way.

We entered Denver at rush hour. As we expected, the freeways were jammed. We decided to test out one of RALPH' capabilities - vision-based platooning. We pulled up close behind a tractor trailer and RALPH locked on. RALPH servoed the wheel to follow the truck, changing lanes and taking exit ramps around the Denver Beltway. During this portion, we covered 36.53 miles with RALPH in control 94.1% of the time.

We pulled off the freeway expecting to have an easy time finding our hotel. Unfortunately, we weren't sure which Embassy Suites hotel we had reservations at. This time Todd's Boy Scout abilities failed us, and we got lost on the surface streets of Denver. We finally found the hotel, grabbed a bite to eat and retired for the night.

REFLECTIONS ON DAY 3

Day 3 was our first real marathon: over 600 miles in a single stretch, nearly all of it autonomous. The early morning power glitch was unnerving — the whole system blinked out, then rebooted as if nothing had happened. Although we brushed it off, it may have been the first hint of the alternator issues that would dog us later. Still, when the system rebooted and kept running, our confidence grew. For much of the day the weather was perfect, the pavement was smooth, and traffic was light — a reminder that RALPH, in the right conditions, could drive almost flawlessly.

Traveling with a reporter from *Business Week* added both pressure and comic relief. The moment when the state trooper ignored our self-driving van and pulled over Otis Port instead was surreal. There we were, a faculty member and his grad student riding in a car with no hands on the wheel, yet the cop zeroed in on the journalist hustling for a photo. It captured the surreal nature of the trip: the future and the present colliding in one absurd roadside encounter.

The stop at Prairie Dog Town became one of the trip's most legendary moments. From the rattlesnakes stirred up for our cameras, to the disappointment of the fake '8,000-pound prairie dog,' to the sad reality of the six-legged steer being bullied by its pen-mates, it was equal parts bizarre and heartbreaking. But it broke the monotony of Kansas wheat fields and gave us a story that still makes me smile. What shocks me now is that Prairie Dog Town kept operating all the way until 2014 before finally closing — a strange little relic that somehow hung on far longer than I ever would have imagined.

By the time we reached Denver, we were pushing RALPH in new ways. The so-called "visual platooning" experiment — where the system locked onto a truck and followed it through traffic — was cool but a little terrifying. The following distance dropped below ten meters, closer than most human drivers would dare. RALPH wasn't truly "seeing" depth; it was simply tracking the contrast between the truck's shadow and the pavement. It wasn't a viable approach long-term, but putting the phrase in our journal was a deliberate poke at PATH, which was heavily promoting platooning with magnets and radar as the future of highway automation. Our tongue-in-cheek version highlighted both the creativity and the craziness of those days.

The day ended with a classic road-trip mistake: getting lost in Denver because we didn't know which Embassy Suites we had booked. After nearly 12 hours on the highway, it was almost funny that navigating surface streets was harder than crossing Kansas.

Day 3 was a microcosm of the entire journey. The technology could shine brilliantly, then stumble in ways that seemed almost slapstick. We chronicled it all in our nightly journal — probably read by only a hundred or two people at most — and we leaned into the humor, sprinkling in jokes and digs alongside the autonomy stats. That mix of real progress and comedy of errors became part of the story. It made the research feel human, not just technical, and it gave anyone following along a reason to smile as we lurched our way into the future.

In hindsight, I think that mix of seriousness and humor in the journal was what made it memorable. We weren't just logging miles and autonomy percentages; we were telling a story, letting the human quirks and absurdities ride alongside the technical progress. That combination is probably why, even decades later, the journal still holds up as more than just a research log — it feels like a snapshot of two people trying to make history, one day at a time.

DAY 4: JULY 26TH, 1995 – DENVER TO GRAND JUNCTION

Driving Time: 2:15pm - 9:30pm

Segment Autonomy Stats:

- *Autonomous Driving Percentage: 95.0% (214.0 / 225.32 miles)*
- *Longest Autonomous Segment: 7.69 miles*
- *Avg Speed: 57.55 mph*

Total Autonomous Stats:

- *Autonomous Driving Percentage: 98.0% (1675.72 / 1709.45 miles)*
- *Longest Autonomous Segment: 68.98 miles*
- *Avg Speed: 61.60 mph*

Our day began at 6:15am, installing Dirk Langer's millimeter wave radar on the roof of Navlab 5 in preparation for an ARPA demo at Lockheed Martin. We drove with Dirk to the site, and collected a bunch of radar images along the way. They looked good, and the radar should eventually allow Navlab 5 to control vehicle speed as well as steer.

We gave a static demo of Navlab 5, and people were quite interested. We sold 10 t-shirts and packed up to leave, needing to make Grand Junction by nightfall. On the way out, we wanted to snap our "Greetings from Denver" picture with the foothills of the Rockies in the background. Unfortunately Lockheed Martin's policy prohibits cameras on site. Undaunted, we snuck off down a side road, pulled to the shoulder, and tried to look inconspicuous until nobody else was in view. When the coast was clear, Dirk raced to the other side of the road and snapped a picture. We think it turned out nicely.

We dropped off Dirk and his radar. As we were leaving, we experienced another power glitch that shut down all computing and actuation. At the time we thought little of it, but later we would look back on this event as a sign of things to come. In about 5 minutes, everything was back on line, and we headed out.

Things were going well, but about 25 miles out of Denver, as we were climbing up into the Rocky Mountains, we lost power again. We unplugged some of our auxiliary equipment (VCR, cellular phone), thinking we might be overloading the vehicle's power system during the climb. It appeared everything was working, so we started off again.

This time, just 2 miles later the power dumped again. We were starting to get concerned. We hypothesized that due to the hot weather, the thermal shutdown on the radio shack inverter that powers the computing equipment was tripping. We needed to directly cool the power distribution box in the rear of Navlab 5, but the nearest air condition vent was 12 feet away on the dashboard. In the spirit of Apollo 13, we unpacked the van and surveyed our supplies. Then it hit us, the giant sun shade we had purchased would just reach from the dash to the power box. We fashioned it into a duct, and cranked the AC. In about 5 minutes, the inverter had cooled sufficiently, so we packed up again and continued.

The power problems caused us to fall 2 hours behind our schedule. We were heading into the mountains with the setting sun directly in front of us. Over

the next 200.32 miles, both RALPH and the vehicle performed admirably. The inverter held up, and RALPH was able to drive 95.0% of the way.

The sun was just falling behind the mesas as we approached Grand Junction, Colorado. When we turned on the headlights, we noticed the battery voltage seemed to drop significantly. We didn't think much of it, as we only had a few more miles to go. Unfortunately, things were more serious than they initially appeared. The voltage continued to drop, and our headlights got dimmer and dimmer. After about 2 miles, the engine conked out, and we were a stealth minivan coasting down the right lane of I-70. We pulled onto the shoulder, lit a flare and considered our options. It was now pitch black.

We tried the cell phone, but there wasn't enough juice left in the battery to power it. We thought our exit was less than a mile away over the next rise, so we started pushing. After about 100 yards, the rise turned out to be more than we had bargained for. We rested for a few moments, pondering the plight of the Apollo 13 astronauts. If they could make it so could we. We remembered that we had the battery Dirk had used to power his radar, as well as some long jumper cables. We quickly cobbled together an Auxiliary Power Unit (APU), and tried the engine. We were afraid the meager charge left in the APU battery would not be enough to start the engine. We held our breath and turned the key. The engine roared to life. After a few quick photos, we loaded the APU onto the passenger floorboard, and secured the connection from APU to the vehicle battery by closing the hood.

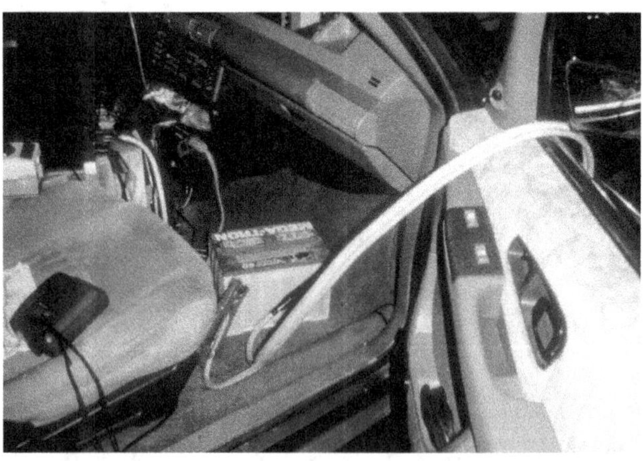

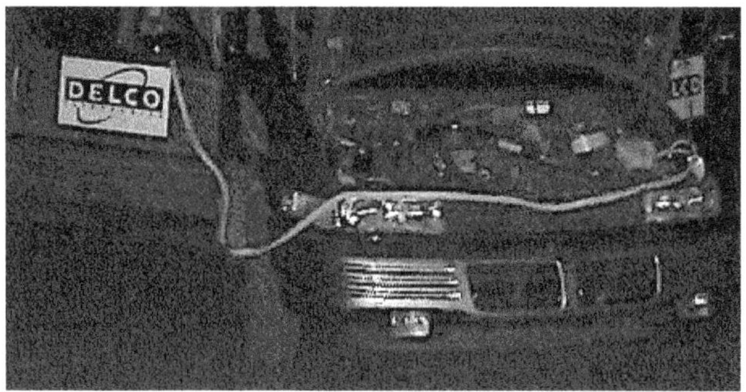

As the electrical engineer on our team, Todd took charge of monitoring the APU. Because of the very limited charge left, we decided to turn on only our hazard lights. To illuminate the road ahead, all we had left was our trusty flashlight. Needless to say, this wasn't quite enough for RALPH and it didn't matter anyway, since we didn't have nearly enough power to fire up the computer. This left driving responsibility to Dean. Unfortunately, Dean had knocked out one of his contact lens' while adjusting the viewfinder on the camera during the hurried photo session. RALPH could have probably done a better job, but both Dean and the APU held up and we limped to the next exit, where we hoped to find a hotel.

We first tried the Comfort Inn, and in keeping with our day's luck, they had no vacancies. Fortunately the Holiday Inn across the street did. We parked, shut off the APU, and went to our room - our most difficult day so far was over.

REFLECTIONS ON DAY 4

This day started with the clean, professional face of research — a radar demo at Martin Marietta (soon to be Lockheed Martin), complete with engineers in pressed shirts and a no-camera policy — and ended with us stumbling into a Holiday Inn after cobbling together our own "auxiliary power unit" from spare parts. Radar and vision were natural complements: radar for obstacle detection and vision for steering. It's striking how that mindset has carried forward — most modern autonomous vehicles still rely on overlapping sensor modalities, combining radar, lidar, and vision. The notable exception, of course, is Tesla, which has staked its future on an all-vision approach, betting that software can compensate for what additional sensors provide.

The mountains tested everything: the van's alternator, our jerry-rigged power system, and our nerves. The AC-duct trick really did feel like Apollo 13, laying a sunshade across the cabin to pipe cool air to the back. When the alternator finally failed completely and the engine died, we coasted down I-70 in the dark, lit only by hazard lights and a flashlight, Dean driving half-blind with one contact lens.

It's crazy to think how fragile the whole enterprise was. A blown alternator or a tripped power inverter could have ended the trip. Yet each time, we found some way forward. That willingness to improvise — to solve the problem with whatever you had in the back of the van — was as much a part of CMU's culture as the research itself. Day 4 reminded us that self-driving cars weren't just an abstract software problem. They were tied to real hardware, real weather, real mountains, and real limits. And when those failed, the researchers had to turn into mechanics, electricians, and problem-solvers on the side of the road.

But it wasn't all stress and breakdowns. We also found ways to make the trip our own — even to put a brand on it. One of the lighter side projects was designing and selling NHAA T-shirts. It was our own version of crowdfunding before that was even a thing. We sold them for $14 apiece and ended up selling about 150 in total. Hardly a financial breakthrough, but they made for great memorabilia and I still have mine. The design itself was simple but effective: the logo on the front, and a map of the cross-country route on the back. Wearing one felt like joining the adventure.

Later in the trip, those shirts took on even more personal meaning. Barb and I wore them for our vow renewal at the Little White Chapel in Las Vegas, standing under a white canopy covered in climbing roses at the famous "Drive-Thru Wedding Window." The photo still makes me smile — two kids in matching shirts, Barb holding flowers, and recommitting ourselves mid-road trip. It was ridiculous, it was perfect, and it was us.

It may sound strange, but those shirts were more than just T-shirts. They were early guerrilla marketing. Branding NHAA with shirts created an identity, a way for outsiders to feel part of the ride, and a sense of belonging for those of us inside the van. Maybe that's why they resonated with me so much. I had grown up in locker rooms, wearing team colors, and those shirts tapped into the same idea: we weren't just researchers hacking code and coaxing hardware down the interstate, we were a team — and teams wear their colors. Later, at the AHS Demo, we'd take the same idea further with trading cards for the vehicles. We didn't call it marketing, but that's exactly what it was — a way to build identity, community, and memory around the work.

DAY 5: JULY 27TH, 1995 – GRAND JUNCTION TO LAS VEGAS

Driving Time: 12:00pm - 9:00pm

Segment Autonomy Stats:

- *Autonomous Driving Percentage: 98.8% (508.16 / 514.18 miles)*
- *Longest Autonomous Segment: 54.65 miles*
- *Avg Speed: 67.50 mph*

Total Autonomous Stats:

- *Autonomous Driving Percentage: 98.2% (2183.88 / 2223.63 miles)*
- *Longest Autonomous Segment: 68.98 miles*
- *Avg Speed: 62.97 mph*

Dean awoke at 6:00am, anxious to work out the previous day's vehicle problems. He called AAA to get a recommendation for an automotive service center. The only AAA approved garage in the Grand Junction area was Metric Automotive. Dean called Metric, and described the problem to Dave, the service manager. Dave agreed that it could be an alternator problem. He said he thought they could fix it today if we brought it in soon. By now, Todd had woken up, and we decided to save time, instead of having Navlab 5 towed to the garage, we would drive the 5 miles from our hotel to Metric using the Auxilary Power Unit.

We made it to the garage without any problem. And in fact for a few moments contemplated forging ahead, since the vehicle battery now appeared healthy. However we quickly thought better of it, since temperature today in Western Colorado and Utah was forecast to be between 100 and 110 degrees, and we didn't relish the thought of breaking down when the closest exit could be up to 50 miles away.

When we told Dave the nature of our vehicle, he looked a little skeptical. We handed him the keys, and he said "check back in a couple hours and I'll let you know if we've had a chance to look at it." We went to the vehicle to get some stuff to work on during the wait, and Dave followed us out of curiosity. After seeing all the equipment in our vehicle, he became noticeably more interested (telling him we were meeting with Jay Leno in Los Angeles didn't hurt either). We left to eat breakfast and check in with CMU.

When we returned an hour and a half later, the receptionist said that Dave and the electrical systems specialist, Keith, needed to talk to us. They had taken Navlab 5 for a test drive, and had determined that it was in fact an alternator problem. They ordered a replacement, and within an hour, it had arrived and Keith had installed it.

For their prompt service and hard work, we rewarded Dave and Keith with complementary "No Hands Across America" tshirts and a live demonstration of the vehicle driving itself. We took a couple of photos, and by noon we were on our way, satisfied customers of Metric Automotive. If you ever break down in Western Colorado, give these guys a call.

We now had a decision to make. We had to be in Las Vegas in 30 hours to meet our wives. The Navlab 5 had already broken down once during a severe mountain climb, and the route we had planned through Four Corners and the Grand Canyon included a traverse that on the map looked nearly as severe as the one from Denver. The original route was longer, and we knew we would need to drive slower. Together these factors led us to decide to stay on I-70 and head directly for Las Vegas. We returned to the sight of our previous night's breakdown and continued our journey.

During the next two hours, the terrain was barren but beautiful. Exits were few and far between - up to 50 miles apart. If this was the beaten path, we were glad we didn't take the less traveled route. During this 139.39 miles stretch, RALPH drove all but 0.91 miles (99.3% of the way).

At the end of this segment, we spied a particularly spectacular overlook, and decided to stop for some pictures. See for yourself.

It was very hot but the new alternator worked perfectly, and unlike yesterday, we rode in air-conditioned comfort. Our next stop was at Burger King/Amoco to get another Pocahontas figurine and fill up with gas. Since it was going to be a quick stop, Dean left the engine running and the keys in the ignition. Todd, in a hurry to get some food, bolted from the car, locking it on his way out. Luckily, Dean had the forethought to crack the driver's side

window. We asked the gas station's auto mechanic for a sharp implement to poke the automatic door lock, and regained access.

Upon entering what we thought was the Burger King, we realized it was a glorified Quicky Mart with a Burger King booth in the back. After waiting 15 minutes for our food, we found out all they had was the Pocahontas statuette - what a disappointment as we already had her. We got her anyway, in hopes of trading her in later for a figurine we didn't have yet. We got the food to go, and before Todd had his seat belt buckled, Dean had finished his chicken sandwich. We pulled back on the highway, destined for Vegas.

We made pretty good time on this stretches of the trip, exceeding the speed limit by too many miles per hour to include in this writeup. Technically, RALPH performed quite well, ignoring the splattered bugs on the windshield, the windshield wiper fluid from the vehicle ahead (during vision-based platooning), tire debris, and specular reflections from the quickly setting sun. There were several long stretch of road without lane markings, and RALPH was able to use the tire wear marks and the road shoulder boundary to continue on. Not wanting to override RALPH during one particularly long autonomous stretch, we tested fate and ran over a rather large piece of retread. The vehicle was unharmed and RALPH continued on to a 54.65 mile uninterrupted segment.

We entered the Glitter City shortly after sunset, and headed for the airport to surprise our lovely wives, who weren't expecting us for another 21 hours.

REFLECTIONS ON DAY 5

Metric Automotive was straight out of the old-school mechanic's handbook: grease on the hands, decades of experience, no-nonsense problem solving. We went in with only a vague suspicion, and they nailed the real issue — the alternator — in a matter of hours. That shop saved the trip.

From there, the decision to bypass the Four Corners and the Grand Canyon was sobering. We wanted the scenic route, but the memory of coasting powerless down I-70 the night before made the choice simple: reliability over romance. The desert is no place for a breakdown.

The gas-station lockout was classic road-trip slapstick. I bolted for food, locked the doors behind me, and stranded us with the engine running. Fortunately, Dean's cracked window saved us from an embarrassing AAA call.

Once back on the road, the trip took on a different feel. With the alternator humming and the AC working, we let RALPH stretch its legs across the Utah and Nevada desert. We pushed faster than we ever had before — well above what we'd normally drive — and RALPH held steady. It shrugged off bugs, windshield spray, even unmarked stretches of road. What impressed me most was how it used subtle tire-wear marks and shoulder boundaries to stay in lane. These faint grooves and oil streaks were barely visible, but RALPH picked them up smoothly — the kind of nuance humans use without thinking, and the sort of thing ALVINN could have struggled with. It was a reminder of just how much progress we'd made.

By the time we rolled into Las Vegas at sunset, the city lights glowing in the distance, there was relief and excitement. Relief that the worst mechanical scare was behind us, and excitement that our wives would join us for the next chapter. Day 5 captured the spirit of the journey — equal parts cutting-edge tech, comedy of errors, and flashes of real promise. The drive had already proven RALPH could go the distance; now it was about sharing the experience with family.

DAY 6: JULY 28TH, 1995 – LAS VEGAS & HOOVER DAM

Day 6, July 28th: We started the day off early, departing the Luxor Resort and Casino to do some sightseeing in the desert around Las Vegas. Before we left, we snapped our customary "Welcome to" photo in front of the Luxor. The person in the long pants and sweat dripping off his forehead in the picture is the esteemed director of CMU's Robotics Institute, Takeo Kanade, who joined us for the final leg of the journey.

Because we missed the second and third biggest attractions of the West, the Grand Canyon and Four Corners (Prairie Dog Town is number 1), we decided that we had to see Hoover Dam. This was the highlight of the day, besides Dean and Terry breaking even on the slot machines.

We toured the facility and then prepared for the first ever autonomous dam crossing. We fired RALPH up and it drove flawlessly across the dam. The

dam had both a yellow center line and a bright red edge line, making it easy for RALPH to traverse. RALPH didn't even flinch at the 500+ foot drop to the bottom of the Black Canyon, but our wives were a little nervous. The cop crossing in the opposite direction didn't seem to notice Dean hanging out the window, flailing his arms wildly as Todd snapped several photos.

REFLECTIONS ON DAY 6

Las Vegas was a milestone in more ways than one. It marked the halfway point of the journey, the reunion with our wives, and the arrival of Takeo Kanade, the director of the Robotics Institute, to witness the final stretch. Having him there lent both pressure and legitimacy. We weren't just two geeks goofing our way across America anymore; the head of the Institute was now in the passenger seat.

Autonomously steering across the Hoover Dam was trivial. The bright lane markings made it easy for RALPH, and the algorithm didn't care if it was on a dam, a bridge, or a highway — to the system, it was all the same. But symbolically, it mattered. A self-driving car gliding across one of America's great engineering marvels was a moment worth capturing.

For Barb and me, there was also the anticipation of what came next:

our vow renewal in Las Vegas. Life wasn't just about algorithms and autonomy; it was about family, commitment, and joy.

The journal captured the day with humor, but in retrospect, this was the calm before the sprint. The desert was behind us, San Diego was within reach, and we were beginning to feel that this impossible trip might actually succeed.

DAY 7: JULY 29TH, 1995 – LAS VEGAS VOW RENEWAL

Day 7, July 29th: Today was the most important day of the trip, at least for Todd and Barb. It was their 5 year wedding anniversary and a big vow renewal ceremony was in the works - complete with a singing Elvis impersonator. Unfortunately, Elvis was too busy and expensive, so they had to settle for a simple drive thru ceremony at "A Little White Chapel." (Joan Collins and Michael Jordan were married at this chapel - not to each other ;) .) Having the ceremony at this location also had the advantage that Navlab 5 could participate in the ceremony as the best "man."

It was 115 degrees out, but the loving couple were cool in their "No Hands Across America" tshirts (Only $14 including shipping.) It was a touching ceremony, despite the fact that the soap bubble gun we purchased at Walmart to lend an air of elegance to the ceremony merely sputtered soapy

water in the direction of the Bride and Groom. The festivities were capped off by the happy couple's exchange of $3.99 rings picked up at the Hoover Dam gift shop. After the ceremony RALPH chauffered the newly re-weds away, past the adult video store next door, dragging beer cans collected from the bushes around the chapel.

REFLECTIONS ON DAY 7

For all the technical milestones of the trip, this day was personal. Barb and I had been married five years, and life had already thrown us into the whirlwind of grad school, research deadlines, and long hours at the lab. But in reality, our story stretched much further back. We had known each other our entire lives and had been dating or married for almost a decade. In many respects, we had grown up together — though she was much more mature than I was.

She was an incredible, caring, kind, smart, beautiful person who I knew would be there at my side no matter what I chose to do or where life would lead us. She put up with my quirks and gave me the freedom to pursue my dreams. More than that, she trusted me to do the right thing for us and for whatever family we would build in the future. That trust was both a gift and a weight. I didn't want to let her down — not just in this project, but in life. To use a football analogy, I had clearly outkicked my coverage.

Barb's Perspective — On Todd's Crazy Projects

> *I never thought Todd was crazy for chasing these projects. I trusted his decisions completely. He always talked things through with me first, so it always felt like a joint choice — not him running off in some wild direction.*

Renewing our vows in Las Vegas wasn't about tradition or grandeur — it was about carving out a moment of joy and commitment in the middle of chaos. The details were absurd and perfect at the same time: the sweltering 115-degree heat, the $3.99 rings from a Hoover Dam gift shop, the soap bubble gun sputtering more water than bubbles. We stood there in matching No Hands Across America T-shirts, laughing through the ceremony while Navlab 5 idled nearby, playing the role of best man. It was ridiculous and heartfelt, just like life often is.

What I remember most wasn't the kitsch, but the grounding it gave me. For all the ambition of the trip, for all the pressure of proving ourselves against competitors like PATH or impressing sponsors and government officials, this moment was just about Barb and me. About remembering why we were doing all this — not just to make history, but to build a life together.

It's fitting that our vow renewal sits in the middle of the No Hands Across America story. It reminded me that even as we chased the future of technology, we were also living our lives in the present, with all the love, humor, and imperfection that entails. And Barb's love, caring, and kindness — then as now — was a huge part of the foundation that made everything else possible.

DAY 8: JULY 30TH, 1995 – LAS VEGAS TO SAN DIEGO

Day 8, July 30th: There was a mixture of sadness and joy in the air, for we knew it would be the end of Navlab 5's epic journey across America. RALPH saved the best for last. Completing the 323.5 mile journey from Las Vegas to San Diego in under 5 hours at an average speed of 70.37 mph, RALPH drove all but 0.51 miles. (99.8%) This included a 56 mile segment, ending when I-15 terminated in San Diego. RALPH wanted to continue to Tijuana, but we had reached our destination.

REFLECTIONS ON DAY 8

The final leg carried a strange mix of emotions. On one hand, we were exhilarated: RALPH had delivered its strongest performance yet, carrying us almost flawlessly from the Nevada desert, through the California highways, and straight into San Diego. On the other hand, there was a bittersweet sense that the adventure was coming to an end.

The statistics told the story — nearly 324 miles at an average of 70 miles per hour, with only half a mile of human intervention. For a car driving using a Sun laptop and an HCII in the back of a Pontiac minivan, with a camera strapped to the windshield and wires everywhere, it was astounding.

I remember joking in the journal about RALPH wanting to push on to Tijuana. That captured the mood: we had proven the point, but part of us wanted to keep going, to ride the wave of momentum further.

As we rolled into San Diego, I thought about how improbable the whole thing had been. Weeks of preparation, AC issues, broken alternators, midnight pushes down dark highways — and yet here we were, with a story no one could deny. No longer was autonomy just a lab demo; it was something that had conquered America's highways.

DAY 9: JULY 31, 1995 – SAN DIEGO TO LOS ANGELES

We woke at 5:30am to complete the final autonomous stretch of the trip - the I-15 HOV lane going into San Diego. He had hoped to drive this 8 mile stretch on the way from Las Vegas the day before, but it was closed for the weekend. We drove the 30 miles to the entrance near Miramar AFB (Top Gun School), fired RALPH up and it performed flawlessly. Like most California freeways, the lane markings weren't very visible, since they were primarily made up of reflectors, but there was a strong edge between the concrete roadway and asphalt shoulder, was well as a distinct oil spot down the lane center. These made it easy for RALPH to drive with only a single interruption as we passed into the shadow of an overpass.

We headed north to Los Angeles, hoping to meet Jay Leno and maybe be on the show. The traffic and smog in LA were awful as we passed "Camp OJ" looking for the NBC studios. After finding the studios and the guest relations office, Todd asked to speak with our contact Helga, at the "Late Show". The receptionist informed Todd the Late Show was on CBS, but they did tape the "Tonight Show" here. She gave us the phone number and we gave Helga a call. After several tries in which she repeatedly put Todd on hold, she spoke with Jay, who was still interested in seeing the vehicle. We waited outside the Tonight Show studio for about 2 hours, until Jay came by heading for rehearsal. He said "hi guys" and walked right past us. Undaunted, we gave Helga another call, and she assured us he'd have time to meet with us during rehearsal.

About 10 minutes later, Helga came by and told us that Jay had gone out to the parking lot to look for us, but couldn't find us. Because she knew how interested he was in the vehicle, she gave us another chance. Takeo and Dean hurried out to Navlab 5 which was parked by Jay's sports car, while Todd went with Helga to get Jay. His enthusiasm was apparent, as he almost ran with Todd out to the parking lot. We gave him a quick tour, and told him what Navlab 5 had done, and he seemed quite impressed. He did say however, that he wouldn't have much use for this technology since he loves to drive so much. We gave him a tshirt, and after a few quick photos, he went back to the studio, and we went to the green room to wait for the show.

The actress who was the voice of Pocahontas in the movie was one of Jay's guests tonight and we thought we might see her while waiting. About one

hour before the show was scheduled to start, she came in, escorted by her entourage. We wanted to take a picture, but thought it might not be cool since we were in the green room, where the "stars" go to relax before the show. She looked very much like the Burger King figurine that had adorned Navlab 5's dashboard since the start of the trip.

Soon after, we were escorted to our seats, which unfortunately weren't on stage. We did get to see the show, but it was clear that the real Pocahontas took precedence over our appearance.

Since we didn't have anything scheduled for the next day, we went to the nearest Mail Boxes Etc. to ship the Navlab 5 computer and electronics back to Pittsburgh. The people there were very nice, and said this was their biggest shipment since they sent a motorcycle to Germany. Everything fit in 6 boxes, and within an hour we were on our way to drop off the vehicle at a nearby hotel, where the shippers would pick it up the next day to haul it back to Pittsburgh.

As we left for the airport, it was a sad sight to see Navlab 5 sitting alone in the parking lot. It had served us faithfully for the past 9 days, driving autonomously 2800 of the 2850 miles across the country. It should get a good rest on its 2 week journey back to Pittsburgh, since it will be riding on a flatbed truck.

We learned a lot from the trip, and there are still areas we need to work on. In particular, the most frequent question we were asked during the trip was about speed control and obstacle avoidance. Over the next year we're going to address these issues, in preparation for our next trip (No Hands Across Asia?).

REFLECTIONS ON DAY 9

The last day began with unfinished business: the HOV lane in San Diego. It may sound trivial, but there was symbolic weight in running those last eight miles. The lane markings were minimal — mostly reflectors — but RALPH adapted, using the oil stains and shoulder edges to stay locked on. It was a small victory, but one that demonstrated adaptability. And in hindsight, it was more than that — it meant we had completed the very first autonomous testing on the same stretch of highway that would become the stage for the Automated Highway System demo just over two years later.

The Los Angeles finale was surreal. Between the traffic, the smog, and the circus-like atmosphere of "Camp OJ," the contrast to Kansas plains or Colorado mountains couldn't have been sharper. Then came the comedy of errors: confusing the "Late Show" with the "Tonight Show," waiting in the parking lot, missing Jay once, then finally catching him on the second try.

Jay Leno's reaction was polite, even enthusiastic — though his line about loving to drive too much to ever use such a technology stuck with me. It was a reminder that public acceptance wouldn't come easily, even if the technical feats were undeniable. And I think it's fun to note that in 2016, Mr. Leno changed his tune and said he was now all for self-piloting cars, even arguing they could help save lives.

Seeing Pocahontas' real voice actress in the green room was another strange twist of fate. For days, the Burger King figurine had been our mascot, duct-taped to the dash, and now the woman behind the character walked right past us in Hollywood. It felt like the universe had a sense of humor.

And then came the end. Shipping off the computers, boxing up the hardware, and leaving Navlab 5 sitting alone in a hotel parking lot was oddly emotional. The van had carried us across America — through storms, breakdowns, endless highways, and moments of triumph. And when I think back now, it's sort of crazy: we had what was probably the most advanced self-driving vehicle in the world at that moment, and we just left it there in a hotel parking lot, the keys tucked on a wheel, waiting for the moving truck to come haul it back to Pittsburgh. We weren't even going to be there — we had a flight to catch. Watching it sit quietly, waiting to be picked up the next day and taken home, felt like saying goodbye to a teammate.

By then, we knew what we had accomplished. Nearly 2,800 miles driven autonomously, coast to coast, at a time when most people couldn't imagine such a thing possible. We also knew the work wasn't finished. Questions about speed control, obstacle avoidance, and broader safety loomed. But No Hands Across America had raised the bar. It showed that autonomy wasn't just a clever research demo — it was a glimpse of the future, proving that self-driving cars could capture the imagination of the world and redefine what was possible on the road.

HUMAN STAKES

But behind the technical milestones, the trip had another layer: the very human stakes of living it, mile after mile.

What really had me worried was the constant uncertainty — never quite knowing the next stop, the next hour, or how the day would unfold. By then, we trusted the system to work or not work. The rest — hotels, timing, meals — always felt a little up in the air, and that gnawed at me more than the algorithms did.

We didn't talk much about "failure" in any greater sense. Neither Barb nor I had any real understanding of what academic projects were supposed to look like. To us, this was research — maybe more public than most, but still only another project. There wasn't outside pressure; the only real pressure was what we put on ourselves: to try something audacious and safely make it work.

The scariest moment came on a dark stretch of Colorado highway, when the alternator failed. Suddenly we were stranded in the pitch black, with no lights and no guarantee we could get rolling again. For a while, it felt like the trip might actually be over. There were lighter, more human moments too — like when we were chatting with a group of Mennonite men in Illinois who, despite living about as far from high tech as you could imagine, were genuinely fascinated by what we were doing. Encounters like that reminded me that this trip wasn't just about computers and code; it was about people trying to imagine a different future.

The final, immediate emotion when we rolled into San Diego wasn't triumph so much as exhaustion and relief. We had made it, and that was enough. Any sense of falling short had nothing to do with reputation or career — it was only about whether we could keep the wheels turning to the end. From our perspective, the upside far outweighed the downside. And if nothing else, the fact that people are still talking about No Hands Across America thirty years later proves the point: it was worth it.

We weren't chasing headlines, but people noticed. What felt like exhaustion, relief, and a string of small daily victories was about to be reframed in the eyes of the world.

MEDIA EXPOSURE

From the moment we left Pittsburgh, the media was watching. Local TV stations followed us, newspapers tracked our progress, and by the time we reached the Midwest, the story had gone national. As mentioned earlier, *Business Week* sent veteran technology writer Otis Port to ride along, and his article, "Look Ma, No Hands" (August 14, 1995), gave our Pontiac van and RALPH a spread in the Science & Technology section. It was strange and gratifying seeing our code described in the same breath as the future of the automobile industry.

Just a week later, *Newsweek* ran Sharon Begley's piece, "The Soul of a New Machine" (August 21, 1995), a forward-looking feature on the future of smart cars. Our trip was right there, side-by-side with Detroit's long-term visions, framed as a taste of what was possible. To have RALPH and our quirky road trip mentioned alongside the R&D budgets of Mercedes, Ford, and GM was both humbling and hilarious.

At the time, NHAA felt like a curiosity — a human-interest story about some researchers doing something wild. But as the decades passed, it was reinterpreted. With the rise of DARPA Grand Challenges, Google's self-driving car project, and eventually Tesla and Waymo, reporters and historians went back to trace the roots. NHAA showed up in Wired, The Verge, NPR, and retrospectives across tech media. What had once been treated as a curiosity was now remembered as a turning point: the first time autonomy left the lab and proved itself on real highways. In that sense, NHAA had two waves of coverage — in 1995, a news blip; by the 2010s, a piece of history.

To me, the media attention was exciting, not intimidating. Back in high school I'd been on TV plenty for basketball, so talking to reporters didn't rattle me. Dean actually did most of the talking — the software was his baby, and he was the natural one to explain it. But when I did get pulled in, I never felt uneasy.

The thing that stood out most to me was Otis Port's *Business Week* article, which focused squarely on No Hands Across America. At the time, that magazine carried real weight, and to see our road trip described in its pages made it feel like our work mattered beyond the lab. Later, other articles would fold our work into broader discussions about the future of the automobile industry, but Otis's piece was different — it was about us, about the trip itself, and it was validation that the effort had been worth it.

Back home in Indiana, the reaction was quieter. I doubt many people beyond our parents even knew what we had done. But later, the local paper ran a story on me — not just about robotics, but about what had become of the high school quarterback and basketball player they remembered. That one meant a lot. My parents were so proud, and it felt like a full-circle moment: the kid who didn't make it to the NFL had still found a way to matter on a bigger stage. For them, pride ran even deeper. They had grown up in a world where telephones in the house were new, and my dad still remembered when the electric company strung lines to their farmhouse. The idea that their son had now helped send a minivan across the country without touching the wheel was almost unimaginable. They were curious, a little in awe, and proud not just of me, but of how far things had come in a single lifetime.

For Dean and me, it was deeply affirming. We had pulled off something bold and new, and we knew it. Pride took a while to settle in — the exhaustion had to fade first — but when it did, it felt good to know that what we had done wasn't just technically sound. It had made people stop and imagine the future.

But in the end, the headlines didn't matter as much as the runs themselves — and the numbers told the real story.

REFLECTIONS ON THE TRIP

When the trip ended and we rolled back into Pittsburgh, the question quickly shifted from how it worked on the road to what it meant. Back at CMU, I think people looked at us a little differently. Dean's ALVINN work had already made him notable globally, but this trip gave his reputation a new dimension: proof that research could leap out of the lab and into the real world. Even though my role was smaller, I felt some of that cachet rub off. It gave me confidence — maybe even the final push to finish my thesis and later lead the AHS demo, which in many ways was NHAA on steroids. I don't think we fully grasped what we had done. We knew it was cool. We knew it was unprecedented. We knew it had worked far better than anyone — even us — might have expected. But the true significance only revealed itself in the years that followed.

By the numbers, the achievement was remarkable: 2,797 miles of the 2,849-mile trip completed autonomously — 98.2%. That wasn't perfection, but it far exceeded what most believed possible. Before 1995, there had been real-world experiments — short highway demos, campus loops, test-track trials — but nothing as bold, public, or coast-to-coast as No Hands Across America. We raised the bar. If you wanted to be relevant, you couldn't stay on the sidelines; you had to take autonomy into the messy, unpredictable real world and prove it at scale.

On a personal level, the trip taught me that research has to be bold. It has to take risks. If we had waited for perfect hardware, perfect algorithms, or perfect conditions, we never would have left Pittsburgh. Instead, we trusted our imperfect system, drove into the unknown, and proved something that has lasted.

It also revealed something fundamental about Carnegie Mellon. We didn't have the most funding or the most institutional polish. PATH in California was buttoned-up and well-supported; Ernst Dickmanns' group in Germany was disciplined and formidable. We were rebels by comparison. But CMU had a culture that attracted people who thought big and valued making things work in the real world. We were researchers by title, systems engineers by necessity — hacking things together until they worked, with the administration backing us to do it. That spirit, more than any particular algorithm, is what made No Hands Across America possible.

A few months later, Dean and I launched our first company, Assistware Technology. The idea was modest — package RALPH as a canonical lane-tracking system and take a small first step toward commercialization. But at that point, Assistware was very much a side project. It ran in the margins of our days: a handful of phone calls, a couple of meetings with people who were curious, and long stretches where it simply sat idle while I finished my Ph.D. and worked on the upcoming AHS project. Still, it planted a seed — one that would sprout in 1998.

The years that followed put the trip in perspective. For a long time, almost no one attempted anything comparable. Not until 2010 did the VisLab Intercontinental Autonomous Challenge send four vehicles from Parma, Italy to Shanghai — nearly 10,000 miles across two continents. In 2015, Delphi drove an autonomous Audi SQ5 from San Francisco to New York, collecting terabytes of data along the way. The technology in both efforts was far more mature, but the spirit was the same: take the system out of its safety zone, put it on real roads, and see what breaks. Looked at in that light, *No Hands Across America* sits at the beginning of that lineage — the original "no hands" run, before the phrase itself even existed.

That single idea — prove it in the wild — rippled outward. Within a decade, DARPA was hosting the Grand Challenge and the Urban Challenge, pushing teams to confront autonomy in deserts, cities, and uncontrolled environments. Today, companies like Waymo and Tesla deploy autonomous systems in dense urban environments. The progression isn't perfectly straight, but the through-line is clear. NHAA forms part of the historical backbone of it all. We showed the problem could be taken out of the lab and into the world, and others carried that challenge forward.

In that context, it's also understandable why *No Hands Across America* receded from the center of the story for a time. The DARPA challenges arrived after the internet boom had reshaped both technology itself and how innovation was covered. The moment favored scale, spectacle, and competition — million-dollar prizes, clear winners, and events designed for headlines. By comparison, what we had done was quieter and harder to package. There was no cash prize and no formal finish line, just two engineers testing whether an autonomous system could truly hold up on public roads. History tends to gravitate toward milestones that are easy to summarize. But the underlying fact never changed. Even three decades later, fully autonomous cross-country drives are still treated as newsworthy achievements. That kind of durability doesn't need defending. It simply confirms what was already true.

In retrospect, maybe it's presumptuous to say that without No Hands Across America there would be no autonomous taxis in San Francisco or Austin today. But I do believe our trip helped create the historical context — a proof point that inspired others to push forward. It showed the world, and ourselves, that autonomy wasn't science fiction anymore. That summer in 1995, in a silver Pontiac minivan held together by duct tape, two researchers, a few allies, and a lot of faith nudged history forward.

And through all of it, it meant a lot that Barb had been in Las Vegas — not just hearing about the work second-hand, but seeing a piece of what I was trying to build. She'd been part of every stage of my life to that point, and it felt right that she had a glimpse of this one too.

VIRTUAL ACTIVE
VISION TOOLS

THE BEAUTY OF A SOLID B

"Perfection is not attainable, but if we chase perfection, we can catch excellence."

— VINCE LOMBARDI

When people hear about No Hands Across America, they naturally assume it was the centerpiece of my graduate work. It wasn't. It was a detour — a thrilling, exhausting, headline-making detour — but a detour all the same. My real job at Carnegie Mellon was simpler and in some ways harder: to graduate.

By then, I had already been at CMU for almost five years. That was still on the short side of what many grad students endured, but to me it felt like an eternity — almost 20% of my life up to that point. And while I had been living what sometimes felt like a fantasy life — playing with cutting-edge toys, doing cool projects, traveling to conferences, even driving across the country in a self-driving car —

there was another, more sobering reality. I was ready to get out into the real world. I wanted to make money, start a family, and move forward with my life. Meanwhile, the world around us was already moving — friends buying houses, settling into careers, starting families. It added a not-so-quiet urgency to finishing. But I knew from the beginning that a Ph.D. was a long-term commitment, and somewhere in the back of my mind I understood that all of those life milestones were going to happen for others while I was still in school. So while it was frustrating at times, I still strongly believed that what I was doing was unusual, that it mattered, and that staying the course would pay off — academically and professionally. In the end, I knew it was the right path.

So while No Hands Across America was exhilarating, the thesis was the thing that really mattered and it was my ticket out. Researching, writing, and defending a Ph.D. thesis was the only way forward, and there came a point when the joy of tinkering had to give way to the discipline of finishing. I began the thesis journey in earnest after my proposal in 1993. Most of my serious work was spread across late 1993, 1994, and early 1995. NHAA occupied the middle of 1995, several months when my thesis was on pause while we pulled off something unforgettable. But when I came back to campus, I knew the fun had to be balanced with focus. The fall of 1995 was devoted to experiments, data, and writing — all leading to a snowy morning in January 1996.

THE BIG IDEA

The thesis was about something I called **Virtual Active Vision Tools**. In plain English, it was about teaching a computer to look at the same road in different ways — without adding new cameras or sensors.

The system at the heart of my thesis was ALVINN — the neural network that had earned its reputation for reliable lane-keeping. It's important to underline, again, that RALPH, the system behind No

Hands Across America, wasn't neural-network–based at all. I returned to ALVINN precisely because it was neural and trainable; I wanted to see whether virtual cameras could stretch its abilities beyond basic lane-following. ALVINN excelled at staying centered in a lane, but anything more — changing lanes, avoiding obstacles, taking exits — was far beyond what it could manage.

My insight was this: if we couldn't add new hardware, we could create new viewpoints virtually. By transforming the images coming in, we could make ALVINN believe it was looking through a different camera. I called these transformations **virtual cameras**, and they gave us a way to manipulate the system's vision without changing the real world. In simple terms, virtual cameras let us give ALVINN "new eyes" without adding new hardware — a mental shift that changed everything.

I can't pinpoint the exact moment before my thesis proposal that the idea clicked. It wasn't a lightning bolt, but rather the recognition that ALVINN was really good at driving on certain road types — yet struggled with tactical driving tasks. We had this great piece of technology, but I wanted to expand it in meaningful ways that actually pushed the envelope of what autonomous systems could do.

Everyone knew those tasks — exit ramps, lane changing, swerving, obstacle avoidance, intersection detection and navigation — were all much harder than simple lane-following. If I could make progress there, it would be a genuine contribution to the field and a justifiable thesis. It also mattered to me personally — proof that I wasn't just along for the ride at CMU, but carving out my own place in the field.

Barb's Perspective — On the Thesis

That thesis took so much of his focus. I knew it mattered to him, and I never doubted he'd finish it, but it felt like it was always in the back of his mind. For me, it was just another season of giving him space to chase something big, while keeping the rest of life steady at home.

GETTING A BIT MORE TECHNICAL

Fair warning: this is where we dip into the technical weeds. Not pages of equations or code, but enough detail to show why it mattered. If you stick with me through the jargon, you'll see how it comes together — and catch one of the neatest parts of the whole story. And if it feels like I'm repeating certain ideas, that's intentional. These are the core principles behind the work — the pieces I most want you to take away — so I'm going to hit them more than once to make sure they're clear.

A virtual camera worked like this: picture yourself standing on a road. Instead of moving your head or picking up the camera, you just used math to tilt, shift, or crop the scene. All of a sudden you're looking a little to the left, or down at the edge of the lane, or way out toward the horizon. To ALVINN, it felt like the world itself had shifted — like I'd just handed it a brand-new perspective. But nothing outside had moved at all. I was only re-imaging the same raw input from the actual camera and tricking the system into thinking it was seeing something new.

That flexibility opened a surprising number of doors. It was as if we had turned a single, fixed camera into a whole array of adjustable ones. By creating these synthetic viewpoints, I could test what would happen if the physical camera had been mounted higher or lower, pitched up or down, or aimed slightly off-center. This formed the foundation for exploring optimal camera placement. You could try them virtually, with math, while still using camera data from the real world.

It also gave us a new way to think about tactical driving decisions. Intersections, forks, or transitions between road types had always been a major challenge for systems like ALVINN. The reason was simple: those situations required symbolic decisions — "turn left here, merge right there" — usually based on hard-coded rules, digital maps, or human operator input. ALVINN, by contrast, was "just" a learning-based system, trained to do one thing and one thing only:

follow the lane in front of it. It had no built-in concept of a road ending, an intersection branching, or the idea of choosing between two valid options.

The real challenge was that there was no natural interface between the two paradigms. On one side were rule-based or deliberative AI systems, which excelled at logic and planning but often slow and lacked robustness in messy real-world conditions.

On the other side were systems like ALVINN — subsymbolic, statistical learners that could grind away at perception problems but had no way to express "choices" in human terms. Trying to merge those two worlds in the early '90s was like forcing two different languages together without a translator. That gap is what made tactical driving so tricky: the symbolic systems wanted decisions, but ALVINN could only give steering angles.

Virtual cameras provided an elegant bridge between ALVINN's world-class lane-following and the higher-level decision-making we wanted to layer on top. By biasing ALVINN's visual input I could guide the system into making what looked like a navigational choice — left or right, this road or that one — without ever modifying its underlying architecture. The learning system simply "saw" a different world, and its trained behavior followed naturally.

When these virtual views were coupled with ALVINN's Image Reconstruction Reliability Estimate (IRRE) confidence metric — a scalar measure of how familiar the current image looked to the network, and thus how much trust to place in its prediction — we could go further. The combination effectively turned ALVINN into a generalized road detector, capable of identifying lane alignments at arbitrary orientations and across diverse road types.

This work was an early hint of what would later become standard in autonomous driving: layered systems where perception modules feed into symbolic planning and decision-making. Most of today's autonomous vehicle companies still rely on that hybrid stack. Tesla is again the notable exception — wagering that a purely end-to-end neural approach, with vision as the sole input and seemingly no explicit symbolic layer, can learn everything needed to drive. And given the pace at which modern AI systems are improving, it's entirely possible that the whole field will eventually converge on fully neural network–based approaches.

Finally, virtual cameras hinted at how to handle higher-speed tactical maneuvers like obstacle avoidance and lane changes. If the view was shifted to the left, ALVINN would steer to the right to re-center the lane — and vice versa. By incrementally nudging the view away from an obstacle in the lane, we could coax the vehicle to steer around it, as if it were making its own evasive maneuver. The same trick applied to deliberate lane changes: sliding the virtual view step by step into the neighboring lane naturally pulled the vehicle across without a jarring switch. It wasn't full planning, but it showed how a simple visual shift could turn a pure lane-follower into something that looked like a tactical driver.

In other words, the real power of virtual cameras was in bridging two worlds of AI that had rarely touched: symbolic reasoning and learned perception. Traditional systems handled these situations with brittle, hand-coded logic. Virtual cameras offered a cleaner interface — letting ALVINN's learned behavior act like a tactical decision-maker without rewriting the system. Virtual cameras created a natural interface between the two — allowing a learned network to take on decisions that had previously been the domain of symbolic logic.

The experiments that followed — on view placement, intersections, lane changes, and obstacle avoidance — showed how virtual cameras could simply and elegantly unify those approaches, extending autonomy far beyond simple lane-keeping. What stood out was how clean the solution was. Most robotics capability improvements of the day involved patchwork fixes, extra sensors, or endless code revisions. Here, nothing had to be bolted on or rewritten — just desired actions shaping geometry, geometry shaping perception, and perception shaping behavior. That clarity is what made it feel less like a work-around and more like a unifying principle.

The next three sections follow that thread: where to look (view placement), how to choose (intersections), and how to maneuver (lane changes and avoidance).

OPTIMAL CAMERA PLACEMENT

A fundamental characteristic of all Virtual Active Vision Tools is that they focus ALVINN's attention on the image regions that matter for the task. In my case, the task was to enable tactical driving behaviors without sacrificing baseline lane-keeping. So the first question I had to answer was simple: do virtual cameras harm lane-keeping — or help it?

To test that, I used a typical real camera mount location on the windshield, collected paired training/test sequences on specific road types, hand-labeled the lane center, and then synthesized many virtual views from the same raw images. For each view, I varied pose parameters — lateral and longitudinal offsets, height, pitch (and, on highway data, horizontal field of view at 30° vs 50°) — trained ALVINN on one sequence, and evaluated on the other. Performance was quantified with the steering position error at a specified look-ahead distance, supplemented by ALVINN's IRRE confidence value and the output sum-squared error.

Results were consistent across road types. There was no single "best" virtual view — many neighboring views produced acceptably low error — yet a few patterns emerged: (1) error tended to increase as the virtual viewpoint moved farther forward, because distant views sometimes saw curvature before the human training driver had begun to steer, creating label mismatch; (2) error often formed a shallow "valley" near centered views, matching the intuition that symmetric road edges/area are the most stable features; and (3) narrower HFOV (30°) generally reduced error versus 50°, especially when spurious features (adjacent vehicles, an emerging exit lane, barriers, shadows) entered the wide view.

In other words, virtual views can do what "structured noise" aims to do during training — exclude distractors — but they do it by cropping them out rather than trying to teach the network to ignore them. In several sequences, the best virtual view even beat the actual camera view (by a few centimeters of peak error), reinforcing that virtual cameras don't degrade lane-keeping and can sometimes improve it.

The takeaway wasn't to chase a mythical perfect physical mount; it was to standardize a good, practical mount and then use virtual tilts/shifts/crops to adapt the view to the road type and situation — keeping near-field features because they stabilize steering, boosting horizon content when curvature prediction matters, and trimming away inconsistent features that confuse the network. That math-driven approach foreshadowed today's multi-camera rigs: we were extracting multiple useful perspectives from one sensor stream, then using error and confidence to pick the right one for the road in front of us.

INTERSECTION DETECTION AND NAVIGATION

One of the hardest challenges for early autonomous driving systems wasn't just lane-following — it was figuring out what to do when the road itself split, merged, or ended. Intersections demanded a different kind of intelligence: the ability to recognize multiple candidate roads, decide which one to take, and then physically steer the vehicle onto that new path without losing control. My work on virtual active vision extended naturally into this space, because it offered a way to look beyond the single forward-facing camera and to create views that tested hypotheses about where roads might be.

Proof of Concept

The first experiments were conducted on Navlab 2, a vehicle equipped with a single fixed camera. The test site was a paved, unlined path on Flagstaff Hill near the Carnegie Mellon campus, about 3.1 meters wide with grass shoulders. The scenario was simple. Could the system, starting 35 meters off the road (in the grass) and pointed perpendicularly at it, detect the road and then navigate onto it?

The approach was to create a virtual camera rotated 90 degrees from the forward heading, positioned about 20 meters in front of the vehicle. As Navlab 2 drove slowly toward the road at 5 mph, the system generated a new virtual view every 0.3 seconds and fed it into ALVINN. Two outputs were monitored: the predicted road position and the IRRE confidence metric described above.

The behavior was clear: when the virtual view imaged only grass, IRRE stayed low. As the view swept onto the road surface, IRRE spiked above 0.8 — well into the range of normal driving confidence. This gave us a simple but powerful rule — a high IRRE response meant a road was present at that virtual view location. The system could then declare that the road had been found.

Detection, however, was only half the problem. Getting onto the road was much harder. The simplest algorithm instructed the vehicle to steer so that its rear axle passed over the detected road center. That worked for position, but not for heading: the car often arrived on top of the road surface misaligned with the road's orientation, unable to continue lane-following.

More advanced methods projected an additional target point down the road (based on the virtual view orientation and ALVINN's output) and used both points to guide a smoother turn. These methods worked in some trials, but they proved brittle — performance depended heavily on parameter tuning (lookahead distance, projection length, detection distance), and the assumptions about local road straightness were often not met.

The conclusion from the Navlab 2 experiments was that while virtual views and IRRE provided a reliable road detector, a fixed camera and dead-reckoning-style traversal were too fragile to handle intersections robustly.

Active Vision for Intersections

The next step was to move to Navlab 5, outfitted with a roof-mounted pan–tilt camera. This active setup solved two problems at once: it expanded the effective field of view (by panning toward likely branch locations), and it allowed continual image acquisition during traversal, reducing the "driving blind" effect of the earlier methods.

With Navlab 5, we could tackle more complex scenarios like Y- and T-intersections. In one set of experiments, the vehicle was told only that an intersection was upcoming, with the approximate orientation of a branch (e.g., 40° left of straight). The system created a Detection View in that direction about 7 meters ahead and panned the camera to ensure overlap between the forward Driving View and the side-looking Detection View.

As soon as the IRRE metric on the Detection View exceeded the threshold, the system declared the branch detected.

In another scenario, the intersection geometry was unknown but the location was given. Here, the system used a radial search: sweeping virtual views around the intersection center at 45° increments. Each view was passed through ALVINN to determine if it imaged a road. Branches imaged in the view produced high IRRE values; spurious directions did not. This allowed the system to map out all the intersection arms, passing the results up to a task-level decision module (or, in experiments, the safety driver) which could select which branch to take.

Localization and Traversal

The harder part was aligning and steering onto the chosen branch. The critical innovation was to use iterative localization with virtual views. This required two steps:

1. **Lateral offset adjustment:** The virtual view that had detected the intersection branch was incrementally shifted perpendicular to the branch until the network's displacement output bracketed zero, meaning the view was centered over the branch.
2. **Orientation alignment:** A second "Projection View" was placed several meters down the branch. If both original and projection views yielded zero displacement, alignment was correct. If not, the angular misalignment was computed and corrected.

This two-step process produced both the lateral and angular correction needed to localize the branch precisely. Once aligned, the system recalculated the branch and required view every few frames, keeping the branch in sight even as the vehicle maneuvered.

Traversal used a geometric construction: treating the vehicle's heading and the branch orientation as tangent constraints, the algorithm computed an arc through the intersection that smoothly connected them. At each new frame, the arc was updated to account for any drift, pan angle adjustments, or nonlinearities in branch geometry. Finally, the real camera was panned throughout the traversal so that it continued to optimally image the road ahead so the appropriate virtual views could be created.

Intersection Detection Results

The system was tested repeatedly on real intersections. Across 35 trials on Y and T junctions, branch detection succeeded in 33. Failures were due to rapid lighting shifts (overcast to sun) that pushed the camera's pixels out of a usable dynamic range — either washing them out or dropping them too dark for the system to handle. In all cases where a branch was detected, the vehicle successfully traversed onto it and continued road following.

The key lesson was that virtual active vision, coupled with confidence metrics and active camera control, could reliably detect, localize, and navigate intersections without any hand-coded geometry models. Instead of programming the rules for every possible intersection type, the system used ALVINN's learned road-following behavior, extended by virtual cameras, to generalize to new use cases.

While the experiments were constrained — low speeds, empty roads, relatively simple geometries — they represented an important proof: intersection handling was possible with learning-based vision systems, years before deep learning became the dominant paradigm. The real contribution was extending ALVINN from passive lane-keeping to the ability to handle decisions at the road level.

Tracking a single straight lane was no longer the ceiling — the system could detect upcoming branches, evaluate them with its confidence metric, and steer onto one, all from a single camera stream and without high-definition maps or modern GPS. It wasn't perfect, but it was reliable enough to prove feasibility: in 1995, we showed that a vision system could do more than just stay between the lines — it could recognize a fork in the road and choose a path forward. That step from following to choosing is the essence of autonomy.

LANE TRANSITION

When people think about lane transitions, the first image that comes to mind is usually the simple act of moving from one lane to the one beside it. But in practice, lane transition covers much more: merging onto highways, taking exits, avoiding obstacles, and even holding a temporary offset within a lane to create buffer space. All these scenarios require more than passive lane-keeping. They require tactical driving behaviors — the ability to form and execute short sequences of actions, in real time, without brittle, pre-programmed plans.

The abandoned Pennsylvania Turnpike near Breezewood was the perfect lab for testing these tactical tasks. Standing there in silence — no cars in sight, just a four-lane highway curving into an empty tunnel — was surreal. It felt absurd and eerie, like a graveyard of interstates. And yet it was also profound: here was a place where we could repeat high-speed robotics experiments with real asphalt, real markings, and no traffic. Hour after hour, I could set up cones, barrels, or boxes, run the same maneuver a dozen times, and measure and log results. That kind of repeatability almost never happened in robotics.

Road Model

To accomplish lane transition-related tasks, I had to create a model that described how multi-lane roads were configured. It was deliberately simple: parallel lanes of constant width and constant separation. This minimal assumption was enough to guide how virtual views were placed laterally, and to keep the system internally consistent. A more complex geometric description wasn't required; the power came from how ALVINN responded when views were shifted. With this model in hand, I developed three techniques for changing lanes.

Incremental Network Switching (INS): The simplest approach was to train two networks — one for each lane — and blend their outputs gradually. Instead of a sudden, unsafe switch from one network to the other, I incrementally re-weighted their steering displacements across ~30 steps. At first, 90% of the decision came from the current lane's network and 10% from the destination. By the end, those percentages had reversed. The result was a controlled, gradual transition that kept lateral acceleration manageable and produced reasonable passenger comfort. Of 38 attempts, 37 succeeded. The method wasn't perfect — transitions tended to "dive" into the destination lane late in the maneuver, a sign of mismatch in how each network handled extreme offsets — however, it proved the blending concept worked.

Incremental View Lane Transition (IVLT): The second technique replaced network blending with view manipulation. Here, ALVINN always ran a single network, but the system created two virtual cameras: a Driving View for control and a Test View, offset slightly laterally.

By checking both the IRRE confidence and the geometric consistency of the Test View, the system could predict whether ALVINN would drive reliably at the next step before actually committing. Once confirmed, the Test View became the new Driving View, and another offset Test View was created. Step by step, the vehicle slid into the next lane. In 32 trials, every transition succeeded, though the method often handed off control before the car was fully centered in the destination lane — a side effect of relaxed constraints in the switching criteria.

Dual View Lane Transition (DVLT): The third method combined the strengths of both approaches by tracking both lanes simultaneously. A Source Lane View (SLV) was placed directly over the center of the current driving lane, while a Destination Lane View (DLV) was offset laterally so that it was centered over the target lane. Each view fed its own trained network, producing an estimate of its lane center — the Lane Center Point (LCP).

The system then steered toward a Modified Lookahead Point (MLP) located between the two LCPs, which slowly moved from one lane toward the other. As the vehicle moved, both the SLV and DLV were updated to stay aligned with their respective lane centers. With each iteration, the system moved the MLP closer to the destination lane, guiding the vehicle smoothly across. Out of 42 trials, every transition succeeded. The DVLT method yielded the smoothest and most symmetric trajectories, with results that felt closest to real human lane changes at highway speeds.

When I got back to the lab and looked through the data, especially from the DVLT runs, a real surge of pride hit me. Not surprise — pride. In the vehicle, the transitions had felt human — smooth, deliberate, almost intuitive — and the plots confirmed it. And the reality was, nobody anywhere had real data on autonomous lane changes at highway speeds — and, frankly, almost nobody could even do it. Maybe one group in the world, maybe, but if they had anything, it certainly wasn't well documented or published. Seeing those plots made me realize something important: this wasn't just a research exercise or a class project anymore. I had built something genuinely new.

Beyond Lane Changes: Obstacle Avoidance and Offset Driving

The same virtual-camera principles extended naturally to obstacle avoidance. By modifying IVLT into the Swerve and Offset Driving Algorithm (SODA), the vehicle could momentarily offset from the lane center, hold that offset to clear an obstacle, and then return. In ~20 tests at 40–60 mph, every swerve maneuver succeeded.

I also tested extended offset driving — holding the vehicle 2–3 meters out of lane center for up to 25 seconds. Though less stable than true lane-following, it was robust enough for scenarios like construction zones or using the shoulder as an emergency escape route in accident scenarios.

Exits, Entrances, and Lane Detection

Lane transitions also underpin highway merging and exiting. To handle ramps, the system combined higher-level cues ("an exit is approaching") with targeted virtual views offset forward and laterally.

Exit detection was robust: in 19 of 20 trials, the vehicle identified the ramp as soon as it appeared in the Exit View and transitioned smoothly onto it.

A similar strategy worked for entrance ramps, merging the vehicle back into the mainline.

Finally, virtual cameras proved useful not only for traversal but for lane presence detection. By combining IRRE confidence with geometric consistency checks, the system could infer whether an adjacent lane actually existed (for example, before executing a lane change maneuver.) This capability extended the system beyond mere reaction, allowing it to reason about the structure of the road.

Lane Transition Results

The experiments at Breezewood demonstrated something important: lane transition, obstacle avoidance, and ramp handling could all be reduced to controlled manipulations of virtual camera views and confidence metrics, without hand-coded models of every case.

Blending networks, shifting virtual views, and tracking lanes in parallel all pointed to the same conclusion: tactical maneuvers could be generalized — different techniques, same underlying principle of using perspective to control behavior. The same framework extended naturally to swerves, offsets, exits, entrances, and lane detection. None of this required rewriting ALVINN or adding new sensors. By manipulating perspective, the system's behavior changed in predictable, useful ways.

The limitations were real: constrained environments, reasonable and consistent road markings, empty roads, and speeds capped a bit below commercial deployment. But the contribution was clear: a robust approach to tactical driving that elegantly unified symbolic and learning based systems. By late 1995, ALVINN was no longer confined to holding a lane; it could shift, merge, exit, and recover — demonstrating, in real time, that vision-based learning systems could take on the tactical maneuvers essential for autonomy.

Looking forward, the echoes are still visible. Most production systems keep a hybrid approach: perception modules and heuristics orchestrate lane changes, merges, and ramps, much like our virtual-view plus confidence framework. Tesla remains the outlier — pushing an end-to-end neural approach where perception and maneuvering are learned together. My bet is that if you could peek inside their network, you'd see it doing, in effect, what we did: reweighting or re-tuning basic road-feature detectors to accomplish tactical tasks.

Even huge end-to-end nets probably can't (and wouldn't) rely on an explicit, hand-coded model of every intersection; they'll learn to modify which visual cues matter for a given moment instead. In short, whether done explicitly with virtual cameras or implicitly inside a monolithic net, the mechanism is the same: change the effective view or feature weighting, check confidence, then commit. That's the crux — with the right views and a trustworthy confidence signal, an extremely capable but niche system like ALVINN could be used to accomplish much more.

THESIS RESULTS

Pulling it all together, the experiments showed that virtual cameras were more than a curiosity — they turned ALVINN from a narrow lane-follower into something closer to a tactical driver, all without adding a single sensor. Knowing I had pushed the system into truly new territory was incredibly satisfying — and even now, I still feel that same sense of accomplishment.

The contemporary landscape was littered with brittle approaches. ROBIN could slide between lanes, but only by "driving blind" for part of the maneuver. PROMOTE's (PROgrammes for MOtorway TEchnology) European prototypes hard-coded lane-change strategies that broke the moment geometry shifted. Even later, non-CMU Automated Highway System Demo vehicles in the U.S. relied on rigid, rule-based control, assuming perfect lane markings and predictable merges. Each worked in its sandbox, but none could generalize.

As for intersections, there really weren't robust detection systems in use then — most groups simply avoided the problem or sidestepped it with maps and predefined geometry. That absence was part of what made my work stand out: ALVINN, with virtual views and a confidence signal, could actually "see" and choose paths at forks where others could not. It was very likely the first real demonstration of robust intersection handling by a vision-based driving system.

In the end, the unifying thread was simple and elegant: desired actions shaping geometry, geometry shaping perception, and perception shaping behavior. That chain turned a lane-keeper into a tactical driver — and hinted at how autonomy itself would one day scale.

To summarize the work and contributions, my thesis showed that autonomy could be stretched far beyond simple lane-following:

- **lane-following** ALVINN already excelled at holding a lane and virtual cameras didn't degrade that capability.
- **Optimal camera placement** revealed that no single mount solved every condition. Low placements excelled in the near field, high placements at curves, but neither worked universally. Virtual cameras bridged the tradeoff, letting one mount act like many and, in some cases, outperform the "real" camera itself.
- **Intersection handling** took things further. With targeted virtual views and ALVINN's IRRE confidence measure, the system could spot and traverse Y-splits and T-junctions. On Navlab 5, a pan-tilt camera kept branches in view, refining orientation mid-traversal — a clean alternative to the brittle dead-reckoning approaches that failed whenever geometry violated their assumptions.
- **Lane transitions** proved highway-speed maneuvers were possible. Incremental Network Switching blended networks to smooth the jump, Incremental View Lane Transition previewed reliability before committing, and Dual View Lane Transition interpolated smoothly between lanes by tracking both centers at once. DVLT in particular felt closest to a human driver's lane change.
- **Obstacle avoidance and offsets** showed how perspective manipulation could masquerade as evasive action. A biased

view shifted ALVINN's perception of the lane, nudging it to steer away from obstacles without ever detecting them explicitly. At 50+ mph, these swerves were stable and repeatable.

- **Exit and entrance ramps** were handled with the same logic. A high-level cue — "an exit is coming" — positioned a virtual view to preview the ramp. If confidence rose, the system committed, shifting control just as in a lane change. Out of 20 exit attempts, 19 worked, with the lone failure explained by training mismatches.

On the abandoned Pennsylvania Turnpike at Breezewood, I had something different: a living lab where experiments could be repeated hour after hour, under real road conditions, but without traffic. Cones, barrels, and boxes stood in for obstacles. Lane changes, exits, and other maneuvers could be run a dozen times a day. That repeatability allowed me to separate luck from real capability.

And Breezewood wasn't the only place — nearly every experiment I ran had a strong real-world component. I tested in Schenley Park, on state roads north of Pittsburgh, and on rural interstates, each location adding a different layer of complexity. In all cases, the work was grounded in actual roads, sensors, and vehicles. That, I think, was one of the real contributions: not just new algorithms, but proof they worked outside the lab. It was also something uniquely Carnegie Mellon — an insistence that robotics live in the world, not just on paper.

What mattered most wasn't the mechanics of lane changes or forks or swerves. It was the principle: by reshaping perception, you could reshape behavior. Without rewriting ALVINN, without bolting on new sensors, the same lane-following system could execute tactical maneuvers simply because its view of the world had been changed.

That was the unifying idea. In a field fractured into brittle, special-purpose hacks, virtual active vision showed that a single architecture could extend upward from lane-keeping to tactical driving. It was geometry, not guesswork. Perspective as control. And it proved, in real time on real roads, that a vision system could do more than hold the line — it could choose a path forward.

THE LAB GRIND

But not every experiment could be done on highways or at Breezewood. That kind of repeatability was priceless, but it wasn't enough. For parts of my thesis, I needed controlled tests that could be run over and over — data my committee could scrutinize in ways you couldn't do with just a video. That meant long days on my office workstation, setting up experiments to churn overnight, and coming back the next morning to sift through the output. I'd repeat the cycle for two straight weeks just to build a dataset big enough to be credible.

And it wasn't just the experiments. To make them possible, I had to build the scaffolding — what today we'd call a test harness. None of it was "official" thesis material, but without it the work wouldn't have existed. I wrote scripts to fire processes in sequence, sweep parameters automatically, log results, and queue the next run. It was infrastructure, not research, but it ate time like nothing else. And like in so many projects, that hidden layer never showed up in the final chapters, even though it was the backbone that made the experiments possible.

This wasn't simulation in the modern sense — no photorealistic worlds, no synthetic data. It was replay, plain and simple: record a stretch of road with a camera, bring the tape back, and feed the same frames through ALVINN in different configurations. Shift the view left, rotate it a few degrees, crop it high, or low. Then do it again. And again. And again.

The contrast couldn't have been sharper. Out at Breezewood, it felt alive — standing on an empty interstate, watching the Navlab 5 change lanes or execute obstacle avoidance swerves at highway speeds. It carried the same energy as our NHAA trip a few months earlier: the sense of being part of something potentially historic. In the lab, it was mind-numbing — hours of silent video loops, rows of numbers, tiny tweaks to virtual views. The grind was incredibly boring, but it was the only way to turn one-off stories into systematic results.

But those weeks of tedium mattered as much as the road tests. Breezewood gave me ground truth; the lab gave me consistency. Together, they made the case: virtual cameras weren't just clever tricks, they were a unifying principle — reshaping perception to reshape behavior, and proving that even a simple lane-follower could be bent into a tactical driver.

I'd done what I came to do — built the systems, run the tests, gathered and analyzed the data. Everything I needed was there. All that was left was the hardest part: turning it into a thesis.

THE WRITING SLOG

If the lab grind tested my patience, the writing grind tested my endurance. Pulling everything together — from Schenley Park fork detections, to driveway transitions north of Pittsburgh, to high-speed obstacle runs at Breezewood — into a coherent thesis was its own marathon.

The hardest part wasn't the experiments themselves but weaving them into a single argument: that Virtual Active Vision Tools were a legitimate step forward in tactical autonomy. I had to show how the pieces fit together — optimal placement analysis feeding into intersection handling, virtual cameras enabling lane changes, Breezewood proving obstacle avoidance — while still keeping the story accessible to readers who weren't buried in the day-to-day details.

I never liked writing, and the thesis didn't change that. I managed to get through it, but it was always a slog. Only later in life would that change — not that I ever grew to love writing, but I eventually became decent at it. In fact, here I am now, decades later, writing this book.

Back then, the tools only made things harder. We used a package called FrameMaker, which was decent for its time but glacial compared to today's cloud-based tools. Integrating images was an adventure — usually done in another program, exported, coaxed into the right format, and then dropped into the document with fingers crossed. Even something as simple as printing in color felt like a challenge designed to break your spirit. I think I only ever made one fully color copy of the thesis; the rest were black-and-white with color pages slipped in manually where needed. I remember hunting down a color copier tucked away in someone's office — a machine I'd never used before — because color printing back then was a genuine luxury item. But I wanted at least one copy of the thesis to look like I imagined it. All of that friction piled onto an already un-fun task, turning the writing process into something far more mentally exhausting than the research ever was.

Draft after draft went through my hands, Chuck's, Dean's, and even Barb's hands. Typos were one thing; logical gaps were another. Every figure had to line up with results, every claim had to be tied to data, and every committee member's likely critique had to be anticipated. It was, to put it bluntly, a royal pain in the ass.

But the grind of writing was just as necessary as the grind of the lab. One gave me consistency, the other gave me clarity. Neither was fun, but both were essential in turning bold experiments into a defensible Ph.D.

THE EQUATION JOKE

After all that grinding on text, figures, and formatting, the math in my thesis turned out to be almost a joke. Literally. My entire dissertation contained only one equation. Only one. And it wasn't some sprawling tensor calculus or a wall of Greek symbols — it was the inverse tangent function, the transfer function for the basic neural networks.

A Carnegie Mellon Ph.D. in Robotics with one equation: even now, it makes me smile. On the one hand, it poked fun at the stereotype of doctoral theses being unreadable math manuals. On the other hand, it said something important about CMU's culture: the value was in proving ideas in the real world, not in how many pages of equations you could stack up. At CMU, building, testing, and showing things work on actual roads mattered more than elegance on paper.

That said, it's not as if there's no math behind it at all. The transformations that turned a real camera's orientation into dozens of virtual views were grounded in linear algebra, projection geometry, and coordinate frame conversions. Once you started shifting the camera around — changing pitch, yaw, offsets, fields of view — the bookkeeping got messy fast. But in all lived inside the code. It powered the experiments, but it never made it onto the page. Which is why the thesis itself could carry the joke of one equation, even though the work behind it was anything but trivial.

THE DEFENSE

All that work led me to one morning: January 11, 1996. The day before my birthday. The day of my Ph.D. defense.

The setting was as unglamorous as it gets: a square, interior conference room in Wean Hall, maybe 25 by 20 feet, with rows of plain desks and those institutional-grade padded chairs. I don't think I'd ever even been in that room before. But on that snowy Pittsburgh morning it was packed. The entire Navlab group turned out, along with others from the Robotics Institute and the School of Computer Science. Despite the weather, it felt like a full house.

My committee sat arrayed before me. Chuck, Dean, and Eric Krotkov took their seats in the front row. Larry Davis joined virtually, patched in through a crackly speakerphone — long before Zoom, armed only with the set of transparencies I had mailed him. And in the very back, quiet but steady, was Barb. She had been with me through every late night, every detour, and every triumph, and now she was there again, for this last step.

One of the strangest things about the day was me. I was wearing a coat and tie. Nobody at CMU wore a coat and tie — least of all me. I don't think I had put one on once during grad school, but I did that day. But only briefly — early in my presentation, I shed the coat, loosened up, and got back into my natural rhythm.

What I remember most is that, somewhat miraculously, every piece of technology worked. That might not sound like much now, but back then mishaps were almost expected. Projector bulbs refused to light, transparency copiers jammed, conference call lines crackled or dropped. But that morning, the bulb lit, the line stayed open, and every transparency came through sharp and legible on the screen. Even my pile of backup slides behaved.

That was part of the strategy. I had deliberately left a few gaps in the presentation, knowing the committee would pounce. Sure enough, the first question — from either Larry or Eric — landed right in one of those gaps. I calmly reached into the stack, pulled out the premade transparency addressing that topic, and dropped it on the projector. That was when I knew the plan had worked. The next couple of questions followed the same script.

Not every exchange was rehearsed. Chuck tossed me a hypothetical — less about the answer itself, more about whether I could think on my feet. I did my best, and his half-smile told me I'd passed that test too. The back-and-forth was tough but fair, exactly what a defense should be.

When the questioning wrapped up, I stepped out while the committee deliberated. The details blur, but I'll never forget being called back in. Chuck, with his trademark dry humor, lobbed me one last curveball before breaking into a grin. The committee, he said, was unanimous — the work was solid, clever, and worthy of a thesis. What it lacked in math, it made up for in showing how to squeeze the most out of the tools at hand — and at Carnegie Mellon that carried real weight.

In that moment, I felt elation — a kind of overwhelming pride that I'd actually done it. I was happy, of course, but it was more than that. For the first time, I felt like I truly, completely belonged. I had contributed something that was mine, start to finish, and earned the six best letters I could imagine: CMU Ph.D.

Barb and I walked out together afterward. I don't remember exactly what we did next — maybe grabbed lunch, maybe just collapsed — but I do remember the mix of relief and anticipation. Relief that this long academic journey was finally over. Anticipating what was already in motion. Assistware was riding the wave of momentum from NHAA. And within weeks, I would also be leaping into the Automated Highway System program, stepping deeper into a faculty-level role that came with real responsibility and national visibility.

If No Hands Across America was the bold, public proof of concept, the defense was its counterweight: the disciplined, defensible proof of contribution. Together, they formed the two halves of my graduate journey. That snowy January day didn't just close the chapter on my Ph.D. — it opened the door to everything that followed.

EUROPEAN VACATION

But before charging into that next storm, Barb and I knew we needed to catch our breath. For the first time in years, there was no looming deadline, no immediate fire to put out. It felt like the perfect moment to pause, reset, and celebrate. So we did something we had never done before: we planned a real vacation, just the two of us. Barb took the lead and found a trip to Europe for us. Neither of us had traveled much at that point. I'd been to Japan once for work, but never overseas for vacation, so the idea of exploring Europe together was incredibly exciting. That trip gave us the space to reconnect and recharge before the intensity of AHS consumed the months and years ahead.

Our flight over was on a wide-body, two-aisle British Airways jet. In 1996, US Airways had a major hub in Pittsburgh and had recently partnered with British Airways to make Europe more accessible. It was funny to think about — just a few decades earlier, US Air had been a small regional carrier serving the Northeast and Mid-Atlantic. When they partnered with British Airways, they even changed their name to US Airways, maybe to sound more international and regal.

We were in normal coach class, but our seats were the two along the outside — just us, side by side. Somehow, on British Airways, that small detail felt different, almost elevated. It was a cool moment, and dare I say, we felt like grown-ups.

We flew overnight, and by the time we landed in London the next morning, I was wiped out from jet lag — barely able to keep my eyes open. Barb had far more energy, and fortunately she pulled me out of the hotel room. That first day, we managed to see Big Ben, walk through Harrods, and even visit Tower Bridge — though I was convinced it was London Bridge.

The next day, we met our tour group. We had signed on with a bus tour company called Globus, and our guide, a very sharp and engaging woman, showed us around London in more detail before leading us to the coast for the next leg of the trip. From there, we boarded a ship for an overnight voyage across the North Sea. Dinner was excellent, with plenty of wine, and later we even found ourselves at a little nightclub onboard. Now, I'm not much of a dancer, but it was fun to be out on the floor with Barb, just enjoying the moment. That whole evening just being together was unforgettable!

Barb's Perspective — On Traveling to Europe

Funny, I honestly don't remember much about the plane trip to Europe, even though I had hardly flown before then. What stands out was seeing all the sights we'd only ever read about in books. I loved walking the streets, sightseeing, and just being there.

The next morning, we reached the Netherlands. We saw a bit of Amsterdam, but what struck me most were the massive windmills scattered across the countryside, along with the famous tulip fields (even though the tulips were no longer in bloom.) It brought back a funny connection: one of the towns in our school district back in southern Indiana was called Holland, and it had its own small windmill. But in the Netherlands, these things were towering giants — functional, iconic, and everywhere.

From there, we continued into Germany, which felt like a homecoming of sorts since both of our families are fully German. We fit right in, enjoying steins of beer in the beer halls, watching people in traditional lederhosen, and taking in the sights. One of the highlights was visiting Neuschwanstein Castle in the Bavarian Alps, the fairytale structure that inspired Disney's Cinderella Castle. Walking its grounds was surreal, like stepping into a storybook.

After Germany, we traveled into Switzerland, where we visited Lake Lucerne and marveled at the scenery. The trip ended in Paris. We did all the classic tourist things — ate croissants that somehow tasted better than any back home, toured the Eiffel Tower, and wandered through the Louvre taking in artwork I'd only ever seen in photo books. One evening, the tour had arranged for us to attend a traditional Paris show. It was fine by us — cabaret-style entertainment with some topless dancing, which, in Europe, was no big deal. The awkward part came when we realized that one of the women on the tour had brought her 8-year-old daughter along. She was understandably upset with the tour guide for not warning her ahead of time, but chalk it up to cultural differences.

Barb's Perspective — On Switzerland and Germany

I especially loved the windmills and being in Germany and Switzerland. It felt fun to fit in there, not standing out so much as tourists.

It also struck us how young we were compared to everyone else on the bus. Other than that young girl, we were easily the youngest couple by a good twenty years. Everyone else was solidly middle-aged. It made us laugh — we stood out without even trying.

Finally, we boarded our return flight at Charles de Gaulle Airport and headed home. It had been an incredible first trip overseas — full of memories, laughter, and the feeling of being out in the world together. Remarkably, it would be another 28 years before we returned to Europe, though that next time was for a very special reason.

One small detail that has stuck with me about this trip was that Barb and I had two seats to ourselves on the British Airways flight to England. It felt symbolic. It was just Barb and me, side by side, carving out our own little space in the world. Later in the trip, on the overnight voyage across the North Sea, it was the same feeling — just the two of us, laughing over dinner, and realizing that these moments of being alone together in the middle of something big and foreign were what made the trip so special.

REFLECTIONS

When I look back on my thesis now, almost thirty years later, I see more than experiments and papers. I see the fingerprints of my upbringing and the values that carried me through. Growing up in Indiana, I was taught to work hard, to be accountable, and to keep moving forward even when the work wasn't glamorous. Those values showed up every time I spent twelve hours on an abandoned stretch of the Pennsylvania Turnpike, or ran image sequences through the system on endless loops in the lab.

My personal philosophy has always been that you don't need to be perfect — you need to be persistent, resourceful, and willing to do the hard, sometimes boring work that others won't. *Perfection is the enemy of success.* The thesis embodied that. It wasn't flashy. It wasn't filled with mathematical pyrotechnics. But it was solid, it was practical, and it worked.

I never thought of my thesis as an "A+ super thesis." I've always said it was a solid B — and I'm proud of that. Because it was an honest B built on real-world experiments, meaningful contributions, and work that held together under scrutiny. I had seen defenses where I wondered how they had been allowed to go forward. That was never the case with mine. Chuck and Dean would never have let me reach that room if they didn't believe in what I had done, and I believe I represented Carnegie Mellon well.

From a research perspective, my thesis showed that you could push beyond lane-following into tactical driving without new hardware. It demonstrated that by manipulating visual input, you could extend its capabilities into harder areas: ramps, lane changes, intersections, and obstacle avoidance. It wasn't the final word on autonomy, but it was a step. And in research, it's the accumulation of steps that matters. And while the dissertation carried only one equation, both the math I learned as an undergrad and the thesis itself taught me the same lesson: structure your thinking, reduce problems to tractable pieces, and design systems that make the most of what they already have.

My work became one layer in the foundation of CMU's legacy in autonomous vehicles — a legacy that would eventually inspire DARPA Grand Challenges, driver support and warning systems, and the very idea of self-driving cars on public roads.

And then there was Barb. In the acknowledgments of my thesis, I wrote:

> *"She has always sacrificed so that I could do what I thought neces-sary to complete this degree. My life would not be complete without her. I love you, Barbie."*

Thirty years later, I can say it still says it best. Her love, patience, and support were the steady foundation under everything else — the long nights, the detours, the experiments, and finally, the finish line.

Finishing my thesis also closed one chapter and opened the next. By early 1996, my formal student journey was done, but opportunity didn't wait. When the chance came to take over the AHS demo, I jumped. I was 28, still young, but grateful that Chuck Thorpe trusted me with that responsibility.

The role itself was a reflection of CMU's unique approach: these were faculty-level positions that didn't necessarily require teaching or doing traditional research in pursuit of tenure. They were funded through **soft money**, meaning you had to write yourself into the projects and justify your place through contribution and execution. That kind of setup was rare at major ROI universities, but it fit the Robotics Institute perfectly — a place where big bets and bold projects took precedence over rigid academic molds.

My thesis was never going to be the flashiest thing I did at CMU. No Hands Across America grabbed more headlines. AHS loomed larger as the next proving ground. But the thesis was the counterweight —

the foundation that let me stand on those bigger stages. Together, NHAA and the thesis captured both sides of CMU's culture: boldness to get out into the real world, and discipline to back it up with proof. The trip showed the world what was possible. The thesis showed the academy how autonomy could be advanced, yet understood and measured.

It gave me the six best letters in the alphabet — CMU Ph.D. — but more than that, it taught me how to think, how to persist, and how to carry forward the values I had been raised with. And in its quiet way, it marked the moment when one chapter ended and another began.

THE AUTOMATED HIGHWAY SYSTEM

THE BIG STEP UP

"The best way to predict the future is to invent it."

— ALAN KAY, COMPUTER SCIENTIST

Walking into the National Automated Highway System Consortium felt a little like showing up to a county fair only to realize everyone else got the memo that it was a black-tie event. I'd never been part of a big company in my life — not growing up in southern Indiana, where every team I was on, whether in sports or school, was small, tight-knit, and built on trust and sweat. Suddenly I was surrounded by Fortune 500 types, government folks in polished shoes, and consultants who could hold a full conversation using nothing but acronyms.

Now, the massive technology challenges didn't scare me — that part felt familiar. I knew what needed to be done, and I trusted the skills I'd built at CMU and my team. But all the other stuff? The people, the paperwork, the administration, the layers of bureaucracy — those were challenges just as big, and nothing in my life had prepared me for any of it. I went from tight huddles and small teams to a world filled with org charts, approval chains, and meetings about meetings. And I had to jump right in whether I understood the rules or not.

It was exhilarating and intimidating and, if I'm honest, a bit absurd. The technical challenges were huge, the political ones even bigger, and half the time I felt like a kid who'd wandered into the wrong huddle. But ready or not, this was the world I was stepping into — and it was going to test me in ways nothing back home ever had.

INSIDE THE NAHSC

The mid-1990s marked a high-water point of U.S. government interest in automated driving. The Intermodal Surface Transportation Efficiency Act (ISTEA) of 1991 set the stage, allocating funding for large-scale experiments in intelligent transportation systems. One marquee mandate was the creation of a National Automated Highway System, with a public demonstration required by 1997. That single clause tucked into a sprawling highway bill gave birth to the National Automated Highway System Consortium (NAHSC) in 1994.

The consortium's core members were a who's who of U.S. industry and research institutions: General Motors, Bechtel Corporation, Caltrans, Carnegie Mellon University, Delco Electronics, Hughes Aircraft, Lockheed Martin, Parsons Brinckerhoff, and the UC Berkeley PATH program. On paper, it was a slightly unnatural but powerhouse coalition, blending automobile manufacturing, aerospace, construction, and research.

Within that consortium, CMU was the smallest player. But I wasn't worried about size or resources — I knew we had the talent and the technical chops to compete with anyone in that room. What concerned me far more was whether our voices would carry. We were the new kids at the table, stepping into a circle of partners who already had years of history together. Even when you're credible, breaking into those long-standing relationships is never easy. I knew we'd have to prove our value quickly, not because we were under-dogs, but because credibility isn't handed out in rooms like that. You earn it.

PATH fielded an established team focused on infrastructure-centric automation — embedding magnets in the roadways to create instru-mented highways as the backbone for full autonomy. It was elegant in theory, but dependent on massive government intervention and investment, since no private entity could retrofit every mile of Amer-ican highways.

CMU's approach was the opposite: vehicle-centric automation, where the intelligence lived inside the cars. Our vehicles carried the sensors, computers, and software onboard, making them capable of operating on ordinary highways. This model could be incrementally deployed, was market-driven, and was far less expensive to the public. Our model was considered evolutionary, while PATH's was dubbed revo-lutionary. Go figure.

That contrast — infrastructure vs. vehicle-centric — framed much of the internal tension within the consortium. You could almost see the lines being drawn: the big consulting houses like Bechtel, Parsons Brinckerhoff, and Caltrans (California's Department of Transporta-tion) favored an infrastructure-heavy model. Embedding magnets in millions of miles of U.S. highways represented a virtual goldmine in construction contracts. Honestly, that approach struck me as a boon-doggle, and not a great use of anyone's time or money.

By contrast, companies like Delco, GM, and Hughes were more philo-sophically aligned with CMU's vehicle-centric approach, but their own bureaucracies kept them from moving nimbly. The notable exception was Ashok Ramaswamy at Delco, the engineer who supported NHAA, who believed deeply in our direction and supported us in ways that mattered. In the end, the split reflected two competing visions: one rooted in government-led construction, the other in market-driven innovation.

An often-overlooked figure in this story was Dick Bishop, the NAHSC program manager at the U.S. Department of Transportation. Managing such a large, diverse consortium was no small task, yet he did it with steadiness and respect. He gave CMU room to deliver in our own way and kept the personalities from derailing the program. After leaving federal service, Bishop went on to a long career as a consultant in automated vehicle systems, carrying forward many of the lessons first forged in the NAHSC years.

When people ask me what the budget for the Automated Highway System program was, I usually have to shrug. The truth is, the numbers were never fully transparent, even to those of us inside the consortium. Officially, the government covered about 80% of the costs, with the member organizations expected to contribute the remaining 20% in either cash or in-kind support. In practice, that "cost sharing" often meant things like Caltrans granting the consortium members free access to the HOV lanes in San Diego, GM donating the Bonnevilles, or Houston Metro loaning us their two transit buses. It wasn't all about writing checks — much of it came in the form of access, equipment, and favors.

From what I've been able to piece together, the NAHSC as a whole operated on an annual budget of roughly $20 million per year, approved one year at a time. However, the full $140 million authorized in the 1991 legislation was never actually obligated. In practical terms, the program lived in the tens of millions, but the detailed allocations by partner were never published, or at least I don't have access to them. I also don't know exactly what CMU's share was, but given our scope of work and the fact that we fielded five vehicles, it was probably in the mid- to high-single-digit millions over the course of five years. Enough to feel enormous compared to No Hands Across America's $20,000 shoestring, but still only a rounding error compared with what Bechtel or GM were receiving. It was a reminder that even in big government programs, the smallest players could still make the biggest splash.

San Diego was selected as the site for the mandated 1997 technology demonstration. That location was chosen not only for having a good demo site on the I-15 HOV lanes north of the city, but also for political reasons. California had strong representation in the consortium and I suppose the federal government, and Caltrans' involvement came with clear financial leverage. By staging the demo there, the program secured additional support and kept one of its most influential stakeholders happy.

MY ROLE

Fresh off finishing my thesis, Chuck promoted me to a postdoctoral researcher. For me at that time, it was a gigantic raise — but it came with a tremendous responsibility. I was put in charge of CMU's portion of the Automated Highway System demo, scheduled for just 18 months later.

In that moment, I felt a rush of excitement more than anything. I genuinely believed we could handle the technical side — maybe that was confidence, maybe a little naïveté. I had no real sense yet of the bureaucracy, the oversight, and the layers of management that came with a project of this scale. In my head, it would be just like No Hands Across America or finishing my thesis: do good work, hit the deadline, and everything would fall into place.

When Chuck handed me the assignment, he offered just one piece of advice. With a grin, he said, "Don't get pregnant." It was his way of acknowledging that this project was going to consume me, body, mind, and soul. What he didn't know was that Barb and I had been talking seriously about starting a family right around that time. I didn't tell her about that comment until much later — but in that moment, it hit closer to home than he realized.

He was right, of course. For the next year and a half, AHS would become nearly my entire life.

The scale of responsibility was night and day compared to No Hands Across America. That cross-country trip ran on a shoestring. CMU's AHS demo project had a budget of roughly $2.25 million, and while I wasn't the only one minding the books, I was expected to make sure our piece of the program didn't go over. That alone was a leap in scope: from worrying about gas receipts and motels to managing line items in the millions.

After six months, the workload was so heavy that I was promoted again — this time into a faculty role as a Systems Scientist. The pay bump was welcome, but so was the reality check: I'd just been dropped straight into full academic adulting. Suddenly I was responsible for deliverables, sponsors, schedules — the whole grown-up package. It was far from the days when my biggest concern was simply getting my code to compile. And once again, I felt like an imposter — capable on paper, but emotionally and organizationally a few steps behind what the title seemed to demand.

Managing the project wasn't just about code anymore. I had to lead a team — and fortunately, it was an excellent one. We had one outstanding hardware engineer, John Kozar, who handled the guts of the computing platforms, wiring, and safety circuits. We also had a few organizers who took on logistics and scheduling. That freed Dean and me to pour our energy into the low-level software and integration.

It was my first taste of real systems-engineering leadership: balancing people, priorities, budgets, and schedules, not just tinkering until something worked. And I'll admit — that part didn't come naturally. I poured almost all my energy into the technical work and far too little into the paperwork, schedules, and reporting that the larger AHS team expected. In hindsight, I can see how much smoother things might have gone if I'd taken those responsibilities more seriously. But at the time, I procrastinated on anything that wasn't directly tied to building the system. I did the bare minimum and kept my routine the same as it had always been: focus on the technology, make it work, and assume everything else would fall into place

The contrast with NHAA couldn't have been sharper. Back then, it was just me, Dean, and a van — a true adventure with no oversight beyond keeping the van running. AHS was the opposite: millions in funding, a hard deadline, and Washington watching. NHAA had been about proving something wild could be done; AHS was about showing the government, industry, and the public that autonomy could work in a more structured, repeatable way.

I was also the face of CMU's demo effort inside the broader consortium. That meant sitting across the table from people at GM, Parsons Brinckerhoff, PATH, and Caltrans, and defending our approach. To be honest, that didn't always bring out the best in me. I've never been great at reading other people's feelings — especially back then — so any perceived condescension hit me hard. And when I thought someone was dismissing CMU unfairly, I got mad.

Sometimes my personality got the better of me. But that fire also fueled me. I loved being in charge of our part of the project, and my goal was simple: make our demo the best. To win. Not just better than PATH, not just good enough for the paperwork, but memorable. Something the public — and the future — would look back on as proof that autonomy could really work.

One flashpoint came with a senior GM manager who was technically in charge of the entire project. He was probably in his late 50s or early 60s, nearing retirement, and I'm sure he didn't want to hear what some young punk like me had to say. But what he was asking of us was flat-out unrealistic, and I pushed back — hard. It escalated into a verbal argument in front of others. In the end, Chuck, and the demo program manager, Terry Quinlan, brokered a kind of détente: I would report directly to Terry, and she would insulate me from further run-ins.

From then on, things went much smoother. The funny part was what I heard afterward. People from other organizations quietly told me I'd been right — maybe not in how I said it, but in substance. They admitted they wished they could speak up the same way but couldn't, stuck as they were in bureaucratic jobs. Hearing that brought a real sense of relief. It made me feel a little less embarrassed about my outburst because it vindicated what I'd been trying to say. I also knew I needed to work on how I delivered things — but to me, that still felt like a step forward.

Chuck, to his credit, had the patience to let me fight my way through — just like he had back in grad school. He'd seen me struggle, learn, and grow enough times to believe I'd do it again. He trusted me more than I trusted myself some days, and that steady faith meant more than I realized at the time. In the end, he believed the demo would deliver because he believed my team would deliver.

Another new wrinkle was the paperwork. Every week I had to file reports to Terry and the rest of the demo team: what we were doing, how we were doing it, and whether we were on track. They wanted development plans, timelines, even a detailed outline of what we'd show in San Diego months in advance. That was crazy to me. I'm pretty sure I rolled my eyes — if not literally, then to myself. I knew every plan and timeline was going to change, probably more than once. Trying to map out unpredictable research months ahead felt like pretending innovation followed project plans. To me, it seemed like a giant waste of time.

Research doesn't work that way — and let's not kid ourselves, that's what we were doing. In a thesis, you don't know the final result until you've fought through the experiments. But here, there was a hard deadline, and I was committed to meeting it no matter what. I understood the need for planning, even if I didn't always love it. So I gave the managers what they wanted to see. And when we finally rolled into San Diego, the sequence of our demo was set up a little differently — not because I'd lied, but because we'd optimized it to show the tech at its best.

From a personal perspective, I think that my role in AHS was the first time I truly felt like a leader since high school sports. It wasn't just me coding late into the night or chasing down a crazy idea with Dean. It was a real project, with budgets, schedules, managers, politics, and a team that looked to me for direction. I made mistakes, I lost my temper, and I bent the rules — but we delivered. And that, more than anything, marked the turning point from grad student to someone ready to lead in the wider world.

And then, right on cue, life decided to have a little fun with Chuck's advice.

WE GOT PREGNANT

Not long after finishing my thesis, Barb and I decided it was time to start thinking about a family. Which, of course, made Chuck's grin and his "Don't get pregnant" warning all the more ironic. Life clearly hadn't read his memo.

Barb did all the research. Now, you have to understand — I was an only child. I had no sisters, and the only thing I knew about pregnancy was what they taught in ninth-grade health class. Barb, in contrast, had read everything. She knew the exact timing and all the little details that gave us the best chance of success.

And it worked. Within just a month or two of trying, we conceived Emma. I'll never forget the moment. It must have been Labor Day weekend. I was still lying in bed when Barb came out of the bathroom holding the pregnancy test. It was positive. We hugged and started crying and laughing right there in the bedroom of our first house. It was one of those moments that instantly changes your life.

Barb's Perspective — On Finding Out She was Pregnant

I probably bought the pregnancy test at the CVS up the street. I was beyond excited when it came back positive! We had waited to finish school before starting a family, and I felt lucky it happened right away. I don't remember what I ate during pregnancy — just that I sometimes needed crackers when I felt nauseous. Thankfully, I never actually threw up.

On the lighter side, the one thing I did know about pregnancy back then was that women were supposed to take folic acid. I had no idea what folic acid was or what it did, but I knew Barb was diligent about taking it — and a lot of it. Evidently it worked, because nine months later Emma was born, a healthy little baby girl.

For Barb, pregnancy seemed — at least from my perspective — relatively easy. I don't really know how to define "easy," but she handled it with a kind of calm and steadiness that amazed me. The only real issue came when she began feeling dizzy, lightheaded, and even had some strange vision problems. The obstetrician sent her for an MRI, which, thankfully, came back fine. They chalked it up to hormones and the fact that Barb was on the smaller side to begin with — meaning things like dehydration and hormonal swings hit her a little harder.

Barb's Perspective — On Doctor Visits and Scares

Dr. Yao was my doctor, and I loved her. I remember once trying to read aloud to my middle school class and suddenly being unable to get the words out, feeling light-headed. I thought I was having a stroke. Going in for the scan afterward was nerve-wracking. I just wanted everything to be okay, especially for the baby. When the results came back fine, I was extremely thankful.

Emma was due May 16th, and that evening Barb woke me up — her contractions had started. I jumped out of bed, threw on a shirt and jeans, and after waiting a bit, we bolted for the hospital. Only later at the hospital did I realize my shirt was on inside out!

The delivery itself was memorable for several reasons. Barb was determined not to use painkillers. Personally, I thought she was crazy — I would have been shouting, "*shoot me up, I'm ready now*" as soon as I walked through the door. But that wasn't her. She wanted to do it naturally. As the contractions grew more intense, Barb eventually decided maybe medication wasn't such a bad idea. But by then, it was too late — she was already so far along that the option was off the table.

One of the wildest moments happened just before it was "go time." Up to that point, the pregnancy suite had felt warm and homey, almost like a nice hotel room. Then, with the flip of a switch, the place transformed. Panels slid open, cupboards popped out, and medical instruments seemed to appear from nowhere. Even a mirror dropped down from the ceiling so Barb could watch the birth if she wanted. It was as if our bedroom had suddenly morphed into a high-tech hospital room. Coming from the robotics world, I appreciated the clever engineering, but it was still surreal to see the room change so completely in seconds.

Then came the hard part. Barb pushed, and pushed, and pushed. Nothing was happening. I stayed at her side, coaching her the best I could, though it was hard to watch her in such pain. The nurses were encouraging, but at one point I overheard one whisper to another, "If she doesn't move this along, we're going to have to take her for a C-section." Barb heard it too — and that was all the motivation she needed. Determined not to go that route, she buckled down with every ounce of energy she had. And then, suddenly, out came baby Emma. I'll say this as delicately as I can: I was at the "business end" and saw everything. Honestly, it was amazing. Since we hadn't found out the baby's sex ahead of time, I looked at Barb at that moment and said, "We have a baby Emma!" We had always known that if it was a girl, that would be her name.

Barb's Perspective — On Emma's Birth

> *Emma was due on the 16th, and I went into labor that day. We stayed home for a while, then left for the hospital once contractions got closer since the drive into the city was long and uncomfortable for me. Labor was tough, but the real fear came from pushing. It felt like things would rip apart, and when I overheard the doctor say I might need a C-section, that was enough motivation to push with everything I had. Seeing our baby girl for the first time was a true miracle. Todd was an amazing coach — very animated as usual, and supportive, vocal, and loving.*

The rest of our hospital stay was smooth, with no complications. A couple of days later, on a humid, rainy afternoon, we brought Emma home. I'll never forget calling my parents with the news. My dad's response was simple but profound: *"I never thought I'd be a grandpa."* That line stuck with me — it was touching, unexpected, and one of those moments I'll always remember.

The first days at home with Emma were pretty uneventful in the big picture, but they were a huge adjustment for Barb. She was largely on her own, caring for a newborn, while I was buried in last-minute preparations for the Automated Highway System demo. Family and relatives came by to visit, but at the end of the day, Barb carried the weight of those first weeks. I only understand now how much strength that took.

Of course, none of this happened in a vacuum. To understand how we ended up juggling so much at once, we need to rewind a couple of months...

BARB'S GOT HER FIRST TEACHING JOB

Not long after we found out Barb was pregnant, she landed her first full-time job teaching middle school kids at Allegheny Valley School

District. She had a 6th or 7th grade class, teaching English and reading. Barb is the kindest, most patient person I've ever met — the last person you'd expect anyone to be cruel to. Her heart had always been with the younger grades, especially first through third, where her warmth and gentleness seemed to fit naturally. But this middle-school job was the opportunity in front of her, and it gave her the start she needed.

And it's important to say: teaching wasn't all hard for her. During her internship at Highlands while she was a grad student at Pitt, she spent part of her time in Marsha Loker's second-grade classroom. Nearly 15 years later, she was at the mall when one of those former students recognized her. He had been just a little boy back then, but now he was grown and working at the Apple Store. He walked up and told her, out of nowhere, how much he appreciated her as a teacher — how patient and kind she had been, and how he still remembered her from second grade. For Barb, that moment meant everything. It was proof that the effort she poured into her students, even early on, had mattered.

But her first full-time job? That was different. Middle school boys are a world unto themselves — loud, restless, and not exactly brimming with maturity. Barb went in every day determined to do her best, but some afternoons she came home in tears. As her pregnancy moved along, the stress cut even deeper. Almost every day brought a new story, and nearly all of them traced back to one boy who made her first year especially painful. Watching her give everything she had and still come home defeated was hard.

There wasn't much I could do except listen, hold her, and remind her she was stronger than the moment. What made it all sting even more was the simple fact that we'd already decided she would take time off after Emma arrived. So every rough day at school felt like stress without purpose — pushing through something she wasn't even going to continue. Meanwhile we were getting ready for Emma, layering anticipation on top of fatigue on top of everything else.

Those months were hard — harder than either of us admitted at the time.

Barb's Perspective — On Teaching

> *After graduating, I worked at Highlands as a reading specialist for a year with grades K-5. It was a one-year position, and then I moved to Springdale High School to teach 7th and 8th graders. I loved many of my students there — though of course there were troublemakers who shot rubber bands or were sassy. My main strategy was to "kill them with kindness." Honestly, it was the only one I had, but it worked.*

All of this was happening as Barb was quickly approaching her due date, and right in the middle of the chaos leading up to the Automated Highway System demo. It felt like life was hitting us from every angle. When I think about it now, the only honest reaction is: what in the world were we thinking? But the truth is, life doesn't wait for the perfect moment. It all just comes — the work, the pressure, the big leaps, the painful days, the joyful ones.

And as it turned out, that whirlwind was pushing us toward the next chapter of our lives — one with more stability, more roots, and a sense of place we didn't even know we were craving.

WE BOUGHT A HOUSE

When we found out Barb was pregnant, everything started shifting. She had just begun her first full-time teaching job, and the reality of becoming parents quickly made it clear we didn't want to bring a baby home to our small row house on Wilkins Avenue. By then I was working as a postdoc, and for the first time in our lives we actually had two real paychecks coming in. After years of scraping by in grad school, that kind of stability felt almost strange — but it also made the decision obvious. As the new year turned in 1997, we finally started looking for a place that felt more like a home, somewhere we could settle in and welcome our new baby into the world.

Our real estate agent recommended we look in the suburbs, so we spent weekend after weekend touring existing houses. Our budget had grown to about $250,000 max, but even then, nothing felt right. We weren't too concerned about school districts at that point — we figured we might move again someday or even leave Pittsburgh altogether — but it was still in the back of our minds.

Finally, after maybe the fifth weekend of looking, and with Barb starting to really show, the real estate agent — perhaps a little exasperated — took us back to her office and simply pulled out the big listing book. Remember, this was before the Internet, so all the houses were printed out on thick sheets of paper bound together like a giant catalog. That's where we found what would become our first real home.

Barb's Perspective — On Finding Our House

I remember looking through that book with you and Patsy, our real estate agent, and just pointing to one house, saying, "What about this one?" The rest was history. When we went to see it, Patsy also mentioned the Kellys (more on them in later books) — a lovely couple we'd need to meet, who were also expecting.

It was a classic starter home: four bedrooms, two and a half baths, an integral garage, in a nice up-and-coming subdivision. The catch was that it sat on the "wrong side of the tracks." Across the road were the bigger, fancier houses, while ours was in the lower-priced section. Don't get me wrong — it was still a huge step up from the row house, and the neighborhood itself remains one of the nicest areas in the northern Pittsburgh suburbs to this day. But the dichotomy couldn't have been clearer.

We went through the buying process, closed on the house, and moved in. Friends from CMU helped us, which wasn't too hard since we didn't own much. The move was quick and easy. Landscaping, however, was another story.

Barb's Perspective — On the South Lake House

> *I remember the house in South Lake so well. It was brand new, and it felt like we had really "made it." It was the perfect home to bring Emma to. I'll never forget Pumpkin, our beautiful Maine Coon cat, greeting us in the garage when we first brought her home.*

I thought I could handle it myself. My dad had always done our land-scaping growing up, and I figured, *how hard could it be?* Well, it turned out to be harder than I expected, and I quickly realized I needed professional help. We interviewed two or three landscapers, and Barb recommended one who seemed reliable but more expensive. I over-ruled her and chose a cheaper option. Big mistake.

That guy was a scammer. I'd give him money, and nothing much would happen until I handed over the next batch of cash. He was impossible to reach, and while we eventually got maybe 75% of what was promised, the biggest problem was the sod. We had ordered sod for the front yard, and though it was delivered, the landscaper never showed up to put it in.

Fortunately, Chuck came to the rescue. He lived in a neighboring suburb, and when I called him in a panic, he came over. Together, the two of us spent the entire day laying sod, racing the clock so it wouldn't die on the pallets. It was backbreaking work, but we pulled it off. I really owed him one for that, and it's something I've always appreciated.

It's hard to believe we packed so much into such a short span of time — expecting our first child, buying a house, and Barb starting her teaching career — all while I was stepping deeper into the Auto-mated Highway System. Life at home was chaotic; life at work was no calmer. And just as we were juggling babies and mortgages, the AHS world was bracing for its own kind of delivery: a demo that wasn't technically head-to-head, but that's exactly how I saw it — a show-down to decide whose vision of the future would win.

THE AHS CONTEXT

The Automated Highway System wasn't just another research project — it was the U.S. government's moonshot for transportation in the 1990s. Congress and the Department of Transportation were worried about traffic deaths, congestion, and fuel costs, and they wanted a bold solution. The vision was audacious: fleets of cars driving them-selves down smart highways at close spacing, safer, and more efficient than human drivers could ever be. To me, that canonical viewpoint had stacked the deck, tilting the entire playing field toward a vision we didn't share. Billions of dollars were at stake, and so was the question of what the future of driving might look like.

I didn't really grasp why Washington was so invested. I knew the talk was about smart highways and tighter spacing, and I knew the money was real, but the bigger picture wasn't what motivated me. For us, it was simpler, almost instinctual: we wanted to win. Our approach was vehicle-centric — put the car on the road and make it work — which by its very nature demanded more software, more integration, and, if we were going to demonstrate it quickly, a fast-and-lean way of building that vision.

That put us in direct contrast not only with PATH, but with what much of the industry seemed to assume was the inevitable path forward: infrastructure-heavy smart highways. What drove us wasn't policy or prestige. It was the rivalry, the challenge, and the determi-nation to prove that our way could work — and work better.

THE TECHNOLOGY

The Automated Highway System demo was less an experiment than a showdown of competing visions. On one side stood PATH — Cali-fornia's Department of Transportation–aligned program — advocating an infrastructure-centric future. Their vehicles traveled in tight formation, guided by magnetic markers embedded in the pavement and by short-range radar looking ahead. In effect, the magnets acted as a buried rail the car could sense, allowing it to pinpoint its lateral position with remarkable accuracy, while the radar measured distance to the preceding vehicle to manage throttle and braking.

It was an elegant, control-theoretic solution — precise, stable, and effective — but only if you were willing to rebuild America's highways to accommodate it. The entire system depended on a highly managed environment where everything behaved predictably. Human-driven vehicles couldn't safely share those lanes; they introduced too much uncertainty and would have disrupted the platoons' tightly controlled spacing and timing. The technology worked, but it represented a top-down philosophy: success depended on massive, government-funded reconstruction on a national scale.

CMU's vision was the opposite: vehicle-centric autonomy. Instead of making smarter pavement, we made smarter vehicles. Our approach relied on sensors, on-board computation, and vehicle-to-vehicle communication when available. It could be deployed incrementally, vehicle by vehicle, without waiting for Washington to rebuild the interstate. And unlike infrastructure-dependent platoons, our vehicles could operate on the same roads as human drivers. In fact, if we did our job well, the person in the next lane would never even know the car beside them was autonomous.

PATH would present an infrastructure-centric vision of the future, while ours placed the intelligence within the vehicles themselves. The demo would allow both approaches to stand on their own merits. And for us, it wasn't a wager — we were confident it was the right direction.

Now that I've laid out the background of the National Automated Highway System Consortium — the players, the politics, and where I fit into all of it — it's time to dive into the system we developed at CMU. As I've said before, it was a technically brutal challenge, and the soft-side frustrations of working with the larger consortium could be exasperating.

But in a strange way, it was exactly what I had spent my whole life preparing for. I had the backing of people at CMU who believed in me, and I had Barb's unwavering trust that I could do anything I set my mind to. She was in her element taking care of Emma, and I knew that no matter how hard it got, she would succeed at that — which meant I could pour my primary focus, for better or worse, into delivering a successful demo.

FIVE PLATFORMS, ONE SYSTEM

One of the defining features of CMU's AHS demonstration was that it wasn't built around a single vehicle at all — it relied on five of them, spread across three entirely different platforms. That alone created one of the hardest technical challenges of the whole project: taking passenger cars, a minivan, and 40-foot transit buses and making them all run the same autonomy stack, behave consistently, and operate as a unified fleet. The CMU "Free Agent" AHS demo vehicles became Navlabs 6 through 10. Navlab 6 and 7 were Pontiac Bonneville passenger cars retrofitted by GM with the latest drive-by-wire technologies. Navlab 8 was a Pontiac Trans Sport minivan, similar to Navlab 5, which we retrofitted ourselves. And Navlab 9 and 10 were 40-foot, 15-ton transit buses provided by Houston Metro and retrofitted by a local Pittsburgh company.

Each vehicle in the CMU fleet carried a carefully engineered sensor and control suite. While the platforms were wildly different in size and dynamics, the hardware payloads were consistent, giving each nearly the same sensing, perception, and actuation capabilities.

Sensing Suite

- **Radar:** Both long-range (up to ~150 m) for adaptive cruise control and obstacle detection and short-range units for side-looking blindspot monitoring.

- **Forward looking camera:** For lane-keeping.
- **Laser:** For monitoring behind the vehicle for tailgaters or fast moving emergency vehicles.
- **Differential GPS (DGPS):** Provided lane-level accuracy, far more precise than consumer GPS of the era.
- **Wheel encoders & inertial sensors:** Added redundancy and smoothed control during GPS dropouts.

Computation & Control

- **Onboard real-time computers:** A single semi-ruggedized Micron Pentium Pro PC running CMU's autonomy stack, connected to vehicle-specific actuator libraries.
- **Custom safety circuitry:** Hardware kill switches, watchdogs, and emergency circuits that let drivers override instantly.
- **Networking:** Vehicle-to-vehicle communication capability (WiFi precursors) — not required for autonomy, but leveraged for cooperative maneuvers.

Actuation

- Full steering, throttle, and brake control on every vehicle (except Navlab 8, which had no brake controller), designed so autonomy could take complete command but safety drivers could instantly retake control.

BUILDING A FOUNDATION THAT WORKED

What really made the multi-platform system possible was how we designed the software. Today, people would call it a modular interface or API. Back then, we just thought of it as clean engineering. Each piece of vehicle actuator interface software used the same function calls with the same names — so the core autonomy code never changed. Instead, we built different libraries that handled the translation to each platform's actuators.

That meant the same autonomy logic could run across a sedan, a minivan, or a 40-foot bus. Add a new vehicle? Just write a new library. This approach gave us flexibility, reliability, and a huge boost in productivity. It was a practical innovation that let a CMU software development team of two people scale to five platforms, and from basic lane-keeping, to full autonomy in less than 18 months.

The scope of the software was enormous for the time: roughly 70,000 lines of C code, written and maintained almost entirely by Dean and me. Between the two of us, we were responsible for 99.9% of the code that actually ran on the vehicles — from the low-level actuator controls, to the core autonomy algorithms, to the high-level user interfaces and even the multimedia systems in the buses.

Dean integrated and extended his RALPH vision system for steering and lane changes. I wrote the obstacle detection and speed control stack from scratch, including what became an early version of an adaptive cruise control (ACC) and situational awareness system. That system integrated forward and side radars, and rearward-looking lasers, fusing them with RALPH's lane position data to determine which obstacles were in our lane and which we could safely ignore. This gave the vehicles an effective 360° sensing bubble — something futuristic then but still the core of how automated vehicles operate today.

On the hardware side, each platform required clever integration:

- **Navlabs 6 & 7 (Pontiac Bonnevilles):** Provided by GM and Delco. They had advanced brake-by-wire and steer-by-wire systems accessible via CAN bus. Speed control was tied into the factory cruise control.
- **Navlab 8 (Pontiac Trans Sport Minivan):** Very similar to Navlab 5, with a steering motor and cruise-control interface for speed control.

- **Navlabs 9 & 10 (Houston Metro Buses):** Each bus used an add-on motor for steering like the Navlab 8. Speed control came through the cruise control system, while braking was handled by a binary valve system: eight hydraulic valves in series, each adding pressure in steps. One valve meant light braking; all eight locked the wheels. You could literally hear the valves clicking in rhythm — mechanical music — whenever the bus slowed down.

All vehicles carried GPS with Coast Guard differential corrections (early DGPS) for close-to-lane-level positioning. They even had an early version of vehicle-to-vehicle communication — essentially Wi-Fi over a serial link rather than Ethernet. It wasn't central to our approach, but it let vehicles share basic situational data.

The system ran on a single, ruggedized, rack-mounted Micron Pentium Pro with custom I/O boards for sensors and actuators. We used QNX, a real-time Unix variant, to keep scheduling predictable. (For reference, in 1996, a top-of-the-line Pentium Pro workstation ran at 200 MHz with maybe 64 MB of RAM. Today's iPhone 16, running at nearly three trillion operations per second with 8 GB of high-speed memory, is more than thirty thousand times faster and holds over a hundred times the memory.) To say the least, by today's standards, it was primitive — but it was rock solid.

John Kozar, our hardware engineer, modified the computing platforms, and built the wiring and safety circuits that "just worked." In an industry where hardware engineers were often late and chaotic, John always delivered early and steady. That gave us confidence in the foundation and freed me to push the software as hard as needed.

This was the peak of my application software coding life. I felt like a true hacker — breaking down impossible problems, fusing sensors, writing controls, and watching the vehicles come alive. The code was clean, scalable, and resilient. In just 18 months and very modest resources, we built the world's first codebase for self-driving cars. It wasn't perfect, but it worked — reliably, across real roads and real weather. It was an extraordinary achievement I'm deeply proud of.

THE DEVELOPMENT AND TESTING PROCESS

Once the hardware and software foundation was in place, the real question was whether it would actually work on the road. Navlab 6 quickly became the workhorse of the group. I drove it almost like my daily car, commuting from my home in the northern suburbs of Pittsburgh to CMU. The interstate doubled as a test track. Every trip had a dual purpose: get me to campus and put new code through its paces. I'd load new software in the morning that I wrote the night before, watch how it behaved during the drive, tweak it at school, and then test again throughout the day and on the way home.

Some days, I didn't even bother going into campus. Instead, I'd stay out on I-79 or I-279 all day, pulling into rest areas or commuter park-and-rides, editing code on the spot, and then heading right back onto the highway to test again. I could do that from nine in the morning until six at night, stopping only for a bathroom break or a quick snack. It was obsessive but effective. Over Christmas break in 1996, with the demo looming, I even drove Navlab 6 home to Indiana so I could keep testing over the holidays. It probably wasn't the best family decision — but at that point, every mile of debugging counted.

Not every test went smoothly. One Sunday around 6 a.m., I headed north on I-79 for my first real attempt at engaging the adaptive cruise control. The moment I switched it on, the car slammed the brakes — hard. We went from about 60 mph to zero in the middle of the highway, tires smoking blue. My heart nearly exploded out of my chest. It could have been disastrous if someone had been behind me. Thankfully, I had checked carefully before engaging. The culprit turned out to be a small software bug — a careless oversight — but the lesson stuck. Testing autonomy on real roads carried real risks. You needed safety nets.

Barb's Perspective — On Risks

I never saw what Todd was doing as too risky. I believed in him, and I had faith it would work out. I don't remember ever feeling scared about our future, even when he was out on the road with experiments or later with startups.

That incident shaped our testing approach. Every system had to have bulletproof fail-safes. On the buses, which would eventually carry thirty passengers, this was especially critical. The first safeguard was that the safety driver always had the last word. If you grabbed the steering wheel or pressed the brake, independent circuits — not our software — instantly killed the autonomy and gave control back to the driver.

We wired in kill switches too, but in practice, the drivers were the fail-safe. We trained them to know what the system could handle, what it couldn't, and what "weird" behavior looked like: a stuttering steering wheel, stop-and-go throttle, or unexplained surges. Vigilance mattered as much as code.

Navlab 8, the Pontiac minivan, became Dean's platform of choice. He used it heavily to refine the vision-based steering and lane-change logic. The buses — Navlabs 9 and 10 — were different. We never tested them on Pittsburgh's roads. They were simply too big, too heavy, and too risky to run in traffic. The only time they operated before San Diego was at the Transportation Research Center (TRC) in Ohio, a closed facility with trained staff who could handle their scale safely.

This was where our modular interface design paid off. We could develop and test the core code on smaller platforms, then drop it onto the buses later with minimal rework. All we had to do was swap in the actuator libraries for the bus hardware and recalibrate the sensors for the larger geometry. Development never had to stop while we waited for bus access. I doubt any other robotics team in the world was attempting to run one software stack across such different platforms. For us, though, it was simply what had to be done — necessity forced the abstraction layers, and they worked.

Now I realize just how far ahead that really was. Even today, most autonomous vehicle companies stick to one class of vehicle to simplify development. As you would expect, Tesla is the exception: they run the same core software stack across everything from Model 3s to the Semi. That's the closest analogy. In 1997, we were already doing the same thing: one brain, many bodies.

Another critical enabler was how we handled builds. We wrote scripts to package up or unpackage the entire codebase from a single tar file, unzip it, and compile and link it. That one command kicked off a chain of scripts compiling everything in the right order. It also saved us from the nightmare of data structure mismatches. More than once, Dean or I would add a parameter to a struct, and the code would compile fine — but the memory alignment would be off, causing spectacular crashes. A clean rebuild always fixed it.

It might not sound glamorous, but this automation was essential. It let us move quickly between vehicles, drop in a new build, and get back on the road in minutes instead of hours. With bus time especially limited, that efficiency mattered. By today's standards it was primitive — basically a DIY version of continuous integration — but the principle was the same: automate the boring, error-prone steps so you could focus on solving the real problems.

What amazed me during this phase was how seamlessly Dean and I worked together. After years of collaboration, we'd developed a kind of shorthand. I'd be buried in sensor fusion and speed control while he was knee-deep in vision. When we merged our work, it usually just clicked.

At the heart of our collaboration was the core problem of automated driving: merging lateral control (steering) with longitudinal control (speed). Dean's vision system gave us precise road position, while my obstacle detection and ACC tagged objects as in-lane, on the shoulder, or in the next lane. Bringing those together enabled safe lane changes, obstacle avoidance, and reacting to emergency vehicles. In many ways, it was the technical expression of how we worked as a team — his system seeing the road, mine interpreting how to move on it. And that fusion became one of the first real steps toward integrated driver-assistance systems as we know them today.

And it wasn't just technical alignment — it was also competitive drive. Dean was a former collegiate swimmer, fiercely competitive by nature. He carried that edge into research, and it matched my own. We both believed the free-agent demo was the future, and we both saw PATH's magnet-heavy approach as the rival to beat.

With Dean pushing vision, John Kozar keeping the hardware rock-solid, and me driving the sensing and integration code, the CMU effort had a rhythm. Independent, motivated, and hungry — and together, stronger than just the basic manpower count.

TESTING AT THE TRC

Every couple of months, we'd pack up the fleet and haul it to the Transportation Research Center (TRC) in Ohio. Those trips were the closest thing we had to formal validation. Unlike the highways around Pittsburgh, TRC gave us controlled conditions: a 7.5-mile closed oval, a massive skid pad, trained staff, and the ability to safely perform high-speed maneuvers, obstacle tests, and multi-vehicle scenarios. It was Breezewood with resources. For Navlabs 6, 7, and 8, TRC meant pushing far beyond everyday Pittsburgh testing. For the buses, it was the only place they could be exercised at all — and once we had them there, we tested everything, relentlessly.

But going to TRC wasn't just about testing — it was about logistics. We learned quickly that the only safe rule was to take everything: spare sensors, extra wiring harnesses, antenna mounts, fake obstacles, and all the infrastructure we might possibly need to simulate the demo. Once we were there, it wasn't like we could run to RadioShack and grab a part.

The track was closed, and we had it largely to ourselves, which meant we could push the systems harder than anywhere else. If an obstacle-avoidance run went sideways and we clipped a barrel, it wasn't the end of the world. (For the record, we never did.) We could also run the vehicles nose-to-tail at gap distances that would've been reckless on a public highway, because we knew there was room to maneuver and no outside traffic.

And logistics didn't stop with equipment. They extended to people and planning. Every vehicle needed a driver, every driver needed to know their role, and every test had to be scripted in advance. We didn't want to waste trips to TRC doing coding marathons; the point was to debug, not develop. To make the most of the track, the vehicles had to be ready, the runs carefully sequenced, and the team organized so that when one test ended, the next could begin with minimal delay.

TRC trips weren't cheap. Factoring in staff, travel, equipment, and a month of testing, I'd guess the bill ran close to a quarter million dollars in 1997 — probably closer to a million in today's money. For a small team, it was a huge investment, which made careful planning essential.

The one downside of TRC was its sheer scale. The oval was designed for cars running 140 mph, with long straights and high banking. Once our vehicles disappeared over the horizon, they were gone for eight to ten minutes before looping back into view. That forced discipline. We had to design tests so drivers weren't just sitting idle, and we had to maximize every cycle of those long laps.

But the size was also its biggest advantage. TRC let us rack up miles like nowhere else. Yes, it was going in circles, and yes, it was highly controlled — but that was the point. We could accumulate hundreds of miles in a single day, all in safe conditions with garages and support staff close by. A typical session might look like this: run 30 miles, pull back into the garage, tweak code or adjust sensors, then go right back out for another 30 miles. Minimal downtime, maximum data.

The skid pad was equally valuable. That giant expanse of flat asphalt became our playground for obstacle detection and avoidance. We could set up barrels, simulate cut-ins, and run evasive maneuvers that would've been impossible — or reckless — on public roads. It gave us the confidence that, come demo time, the vehicles wouldn't just cruise straight lines, they could handle surprises.

TRC was also the only place we ever ran the buses before San Diego. Those massive 40-foot platforms had to be calibrated, tuned, and shaken down in a safe environment, and TRC gave us that. The same autonomy stack that ran on the Bonnevilles and minivan was dropped onto the buses with only library changes. It was living proof that our "one brain, many bodies" design really worked.

Another milestone at TRC was our first real test of primitive vehicle-to-vehicle communications. It wasn't anything fancy — but it was enough to let vehicles share basic data. The setup was simple but profound: a lead vehicle would spot an obstacle and radio its location back to the followers, who were running so close they'd never have time to see it themselves once the leader moved aside. Instead, they executed avoidance maneuvers automatically — swerving around something they never directly observed. Watching two Free Agents coordinate like that on the TRC track was a small but powerful glimpse of the future.

Finally, TRC was where we met the Houston Metro guys, who had loaned us their buses. They arrived half-skeptical, half-curious — and left floored by what they saw. They were flabbergasted it even worked, and at the same time couldn't have been more supportive. Despite differences in age, background, and geography, we clicked immediately. They sensed in us the same Texas streak they carried: not exactly rebellious, but determined to push against convention and do things our own way. And as an institution with looser purse strings, they were the ones who always sprang for pizza and refreshments on their nickel — the kind of thing we could never dream of submitting back at CMU.

Their involvement made sense once you understood their scale. Houston Metro was the third-largest transit system in the U.S. They had their own police force, built parts of their own infrastructure, even elevated bus lanes through the city. They had a track record of embracing new tech, and the AHS Demo was their chance to see what the next generation of buses might look like.

For us, their partnership was gold: not just access to two heavy-duty platforms, but allies who believed in pushing the frontier. Their trust carried through all the way to San Diego, where those buses became two of the stars of the show.

By the time we wrapped our last sessions at TRC, we knew the vehicles, the code, and the people were as ready as they could be. The bugs had been beaten out, the safety nets rehearsed, and the buses shaken down alongside the Bonnevilles and the minivan. What remained was the hardest part: taking it all out to the I-15 HOV lanes in San Diego, and proving it in front of government officials, industry leaders, and the national press.

But if building and testing the system was one kind of challenge, navigating the consortium was another. Before the rubber ever hit California pavement, we had to survive the politics, paper-pushing, and endless scrutiny that came with being the outsiders in a massive national program. That was another very real test.

POLITICS

The bureaucracy was the hardest part. Parsons Brinckerhoff, GM, Bechtel, Lockheed Martin — all huge organizations built on process and paperwork. I wasn't wired for that. I wanted to build, test, and prove things on the road, not spend hours writing reports that no one would ever use. Acting as a liaison between that world and CMU's fast-moving approach was tiring, and I lost my patience more than once.

Still, we had allies. Ashok from Delco backed our vision of putting intelligence into vehicles instead of pavement, and Terry Quinlan gave us room to deliver in our own way. Without her, politics may have sunk us.

What frustrated me most wasn't the people — many were sharp, committed engineers — but the system around them. Endless reviews, layers of approval, and a tendency to protect process over progress made even small decisions slow. At CMU, you solved a problem by rolling up your sleeves. In the consortium, solving a problem often meant filling out a form.

I didn't appreciate it then, but navigating that world was part of the job. It wasn't enough to have better technology; you had to find ways to move it forward inside a structure that wasn't built for speed. It took me years to learn that lesson, but AHS was the beginning of it — understanding that progress often requires engineering around people and process just as much as around code and sensors. But looking back, that was the part of the job that drained me the most — and the one I'd happily never repeat.

THE 1997 AUTOMATED HIGHWAY SYSTEM DEMO

By the summer of 1997, our vehicles had thousands of miles under their belts — endless loops at the TRC, mixed with countless real-world runs around Pittsburgh. Next, the focus shifted to the real stage: I-15 in San Diego. Unlike TRC, this wasn't a closed test facility — it was a live stretch of interstate that had been temporarily shut down and retrofitted with barriers to create a demonstration zone.

Specifically, the demo took place in the HOV lanes, which ran down the middle of the freeway. Jersey barriers lined both sides, keeping the demo vehicles physically separated from regular traffic. That protection was reassuring, but it didn't eliminate the risks. We weren't just running autonomous cars in front of engineers anymore — we were going to be carrying members of the public, VIPs, press, and government officials. Everything had to be extremely polished and, above all, safe.

For me, the prep felt like déjà vu — the same relentless cycle of testing and debugging, only now under California heat and with a newborn in tow. Dean and I went out first, at the end of June, to begin setup. A couple weeks later, on July 14th, Barb, and seven-week-old Emma, joined us. (Bala Kumar, one of my CMU colleagues, even helped Barb wrangle Emma on the plane — a small kindness I'll never forget.)

Barb's Perspective — On Bringing Emma to San Diego

Flying to San Diego with a newborn was nerve-wracking. Emma was so little, and I hadn't flown much myself. Thankfully, Terry and Kyle were on the same flight, even if Terry had her hands full with a toddler. Staying in the condo on the ocean turned out to be wonderful. I could take walks along the beach, and Terry became a steady companion through those long days.

TESTING AT THE DEMO SITE

Before we let a single VIP onto the vehicles, we had to prove everything on the ground. That meant about six weeks of development testing at the San Diego site itself. The conditions couldn't have been more different from Pittsburgh or Ohio. The HOV lanes where the demo would run were inland, surrounded by concrete and scrub, and the heat was relentless — often pushing 100 degrees. Overheating was a constant concern, not just for the vehicles but for the racks of computers and electronics crammed into their trunks. To keep things alive, we often left the cars idling with the air conditioning blasting between runs, just to keep the equipment bays cool.

The first three or four weeks were essentially seven days a week, twelve hours a day, with minimal breaks for lunch. We ate whatever snacks we could scrounge, washed them down with Coke, and just kept going. In the evenings, I'd go home to see Barb and baby Emma, hear what they had done during the day, and then tackle whatever mundane administrative tasks still needed attention. Chuck was a godsend at this point — he largely took over the bulk of the paperwork and coordination with the consortium, which let us stay focused on development and testing. Those weeks were exhausting, and the only reason I survived them was because Barb carried nearly 100% of the family load.

Looking back, I probably should have been a better dad during those first weeks. Or at least a more present one. But the truth — and I'm a little embarrassed to admit it now — is that I was completely absorbed in what I was doing. I put the project first. Some of that was the selfishness that comes from being an only child; some of it was just immaturity. I was still learning that the world wasn't just about me anymore. And yet, even in the middle of all that, I knew Barb was a natural. Being a mom was something she was simply built for, and I trusted her in a way that came from the deepest part of my gut. What I didn't understand then was what I was missing by not being more present. It's a trade-off a lot of fathers make, but for me, I was lucky enough to earn all that time back in spades later on.

Barb's Perspective — My Hardest Moment

If I'm being honest, the hardest stretch of that entire period wasn't the long days or the constant juggling — it was when I got mastitis in San Diego. It hit fast, and I was sick all over: fever, chills, pain that made feeding Emma almost unbearable. And the worst part was feeling so far from home — I was on the other side of the country with a newborn, trying to hold it together.

I don't think I told Todd how bad it really was at the time. He was buried in the demo, and I didn't want to pile anything else onto his plate. But inside, I was scared — not just for myself, but for Emma. When you're a new mom and something goes wrong with your body, all you can think about is how it might affect your baby. That worry sat on top of everything else: the distance, the exhaustion, the feeling of being alone in a place that wasn't home. Looking back, I'm proud of myself for getting through it. But in that moment, it was the closest I came to feeling completely over-whelmed.

Our base at the site was a massive Quonset hut that served as a garage, which at least gave us shade, and Houston Metro added a nice touch by renting a construction trailer nearby. It had blessed air conditioning, a fridge stocked with water and Coke, and just enough desk space for us to huddle over laptops. They weren't just partners — they felt like hosts, making sure we had what we needed so we could focus on the work.

And the work was endless. We spent those early weeks running scenarios in isolation — lane-keeping, cut-ins, merges — and then combining them into the longer sequences that would become the demo. The sheer technical scope of the software was massive: steering and throttle control, mapping, vehicle-to-vehicle comms, sensor fusion, driver displays, and in the buses, a full multimedia system.

Okay, now it's time to come clean. By the time we moved to San Diego, we had all the core technology completed. The vehicles could stay in their lanes and handle basic adaptive cruise control. But everything else — lane changing, obstacle detection and avoidance — was still developmental. And several major pieces, like the mapping system, the demo-sequencing logic, and other integrations, hadn't been written yet. We'd pushed those to the end, figuring we could finish them with the extra time we'd have on-site in San Diego.

I'm absolutely positive I didn't present it that way to the consortium. I made things sound much rosier than they really were, at least by their standards. But by ours, we were exactly where we needed to be. We had knocked out the hard, foundational problems, and we were confident in our ability to pull the rest together. Some might call that risky — maybe even a breach of trust — but that's how we operated. We were engineers. We knew what needed to be solved first and what could wait. There was no way I was going to let a bureaucrat in an office tell me the right order to build a system, or worse, try to cancel parts of the demo because they didn't understand what mattered and what didn't.

Of course, there was risk — the risk it wouldn't work, the risk our assessment was wrong, the risk that betting on ourselves might backfire. But that was a chance I was willing to take, and a chance the team was willing to take with me.

Barb's Perspective — On Staying Grounded Amid Chaos

> *I grew up in a family with eight kids, so chaos never scared me — I thrived in it. The chaos of CMU years never felt overwhelming because I was used to that level of noise, activity, and unpredictability. Growing up in a big family gave me toughness and resilience. It taught me how to get along with others and how to handle chaos without letting it swallow me.*

While Todd was off developing technologies that made my head spin, I just did what I've always done: keep moving, keep steady, and take care of what needed to be taken care of.

One of the things we ended up adding at the demo site — almost on the fly — was a full multimedia system in the buses. I'm not even sure we ever formally told the consortium about it, and to be honest, I don't think we'd even thought about it before arriving in San Diego and realizing how many people would be onboard and how hard it would be for them to see what was happening. Once we saw the scope of the demo, it was obvious: if the audience couldn't follow what the autonomy was doing, the whole demo would fall flat. The Houston Metro guys were all for it — I think they even found a little extra money in the contract to make it happen — and in the end, the system turned out to be really cool.

We mounted TVs inside the buses, driven by custom video switchers tied to small cameras placed inside and out. At predetermined GPS coordinates, the switchers would automatically flip views, synchronized with the script Dean and I read over microphones to the passengers. It wasn't Hollywood, but it came close — a polished production designed to make autonomy feel real and understandable to people who had never seen anything like it before.

The site also taught us discipline — and reinforced just how much we depended on it. We couldn't afford surprises, so we treated every run like the real thing. The more we tested on site, the more it sharpened our focus and tightened our routines. And just to be clear, none of this was 'flying by the seat of our pants.' We understood the gravity of what we were doing. The only reason we could adapt on site — make changes, add features, solve new problems — was because the core autonomy was solid. We had a strong technical foundation that worked: lane-keeping, adaptive cruise, the whole backbone of the system. We trusted it because we had built it, tested it, and knew exactly what would work, what wouldn't, and what to be cautious of. That rigor — the scientific training behind it — is what made everything possible. We weren't guessing; we were leaning on years of methodical work.

Drivers were trained to know the system's limits, and the safety nets were drilled into muscle memory. The disengage options were bullet-proof: grab the wheel or hit the brake, and autonomy was out instantly. No software involved — just hardwired circuits. That safety culture had grown out of hard lessons back in Pittsburgh, and by San Diego, it was second nature. Every rehearsal carried the weight of the demonstration ahead. If something failed, we reset and ran it again; if something worked, we ran it again anyway. By the end, the vehicles had accumulated thousands of miles on that single closed stretch of I-15.

And looking back, that mix of disciplined rigor and a little bit of rogue spirit was exactly who we were — engineers who understood the stakes, believed in the future we were building, and felt, even then, the faint sense that what we were doing might just matter.

For CMU, it marked the culmination of nearly a year and a half of effort, distilled into a few miles of California highway. Secretary of Transportation Rodney Slater would be there, along with nearly everyone who mattered in transportation and automation. The days of isolated tests and quiet victories at TRC were over — what came next were full-on dress rehearsals, the last step before stepping into the spotlight.

DRESS REHEARSALS

The formal dress rehearsals were where practice met pressure. They weren't just box-checking for safety — they were about building confidence, proving to ourselves that the approach would hold under the spotlight. We knew our method was right. We knew it was going to work. And the more we ran through the script, the more that confidence showed.

The systems were tuned, the scripts polished, and the team worn down to a sharp edge — but at its core, this was still barely more than research. We knew anything could happen, and we'd have to handle it on the fly. And with the bureaucracy of AHS overhead, every tweak or adjustment came under a microscope. When the Demo started, there would be no hiding and no second chances — it had to work, and it had to work safely, with the world watching.

One of the hidden challenges was automating the startup procedures. These weren't just cars you switched on and drove; they were rolling research platforms that had to be coaxed into life in exactly the right order. Because GPS in 1997 was still coarse and unreliable, it often took time to lock into a position precise enough to trigger the autonomy stack. We spent hours fine-tuning the timing so that when a vehicle pulled out for a run, everything came alive right on cue.

To keep coordination tight, we leaned on simple voice radios. The team worked out a lightweight protocol so the chatter was minimal once a run began. Everyone knew their role, and if one vehicle hiccupped in a non-catastrophic way, the protocol was clear: the driver took over manually and stayed in formation, while the rest of the vehicles continued autonomously. For example, if one of the buses failed to pick up its GPS lock at startup, the driver would simply keep pace under manual control while the other vehicles completed the scenario. That way the run still looked seamless to passengers, and the overall demo didn't grind to a halt.

And when hiccups did happen during rehearsals, we used engineering judgment. We were the most equipped to decide whether something was a big issue or not — and usually, it wasn't. We were also the calmest about it. The higher-ups were under enormous pressure, but they had no real control over what happened on the vehicles. They were the fluff; we were the substance — or at least that's how it felt to us. So there was no use getting them spun up over every minor glitch and making our lives harder. We'd note the issue, fix it later, and by the next run, it would be sorted.

There were a few lighter moments too. The site was near Miramar Air Force Base, home of the real Top Gun school. We never saw the kind of acrobatics from the movie, but every so often jets would roar overhead, a reminder that we weren't the only ones testing high-performance systems in the California heat.

By the time rehearsals wrapped, the script wasn't just practiced — it was ingrained. Systems, drivers, and teams knew their parts. From the outside, it looked seamless; from the inside, it was the product of weeks of grinding discipline, long hours, and a safety culture forged through experience.

As the last rehearsal run finished, I just stood there watching, taking it all in. Five vehicles — all running software I'd helped write — moved smoothly along that stretch of highway, reliable and steady, ready to be shown to thousands of people. It was incredibly cool. And at that moment, it hit me: this wasn't research anymore. I was looking at the future — one I'd helped build — running right there on the road in front of me.

People often ask me how it felt to watch it all come together, and the truth is it came in layers. First was the enormous sense of personal pride in the technical work I'd done. But just as strong was the pride I felt in the team. They had endured the long hours right alongside me, and I hoped they understood the magnitude of what they helped accomplish.

And then there was the excitement — not just about what we were about to show the world or the possibility of "winning" in the eyes of the technical community and the deployment people, but about the unknown. There was this sense that what we were doing might open doors none of us had imagined yet. That thought hit me too, at least a little, and by then I was mature enough to recognize it.

As we geared up for Demo Day, I thought back to No Hands Across America. That trip had been scrappy, improvised, and fueled as much by T-shirts and stubbornness as by algorithms. AHS was bigger, polished, and backed by the Department of Transportation, but in many ways the spirit was the same. Both were about putting something real on the road and letting the world react. NHAA had been a spark, a curiosity that made people wonder. AHS was about to be the spotlight, designed to show the world — in no uncertain terms — what the future might look like. And this time, the world wasn't just curious — it was watching closely, with cameras rolling and headlines waiting to be written.

THE INCIDENT AND THE RIVALRY

PATH, our primary counterpart, continued to bet everything on their infrastructure-centric magnet-and-radar system — elegant on paper, but dependent on perfect pavement and tightly controlled conditions.

During one of their prep runs, PATH suffered a rear-end collision. Nobody was hurt, but it exposed the fragility of their method. To be honest, we were secretly a bit happy it happened. Not because we wanted anyone hurt — we didn't — but because it punctured the big claims they'd made about the strength of their control theory. Complex systems rarely fail cleanly; they fail in complex ways. And when the human has absolutely no chance to intervene, that's a recipe for disaster. If their system had been deployed and a failure had been systemic, the hundreds of millions of dollars invested in magnets, pavement modifications, and specialized vehicles could have been wiped out overnight.

For us, the incident only reinforced what we already believed: the vehicle-centric, sensor-driven Free Agent model was the only path to eventual deployment. The rivalry wasn't personal — it was motivational, and it was grounded in conviction. If this technology was ever going to see the real world, it had to be introduced incrementally, layered onto ordinary vehicles and everyday roads.

And yet, I want to be clear: I respected the PATH team immensely. Steve Shladover, Wei-Bin Zhang, Jim Misner — they were sharp, dedicated, and serious about their work. My rivalry with them wasn't about personalities; it was about approaches. They advanced control theory in tightly constrained environments, and I'm sure what they developed remains valuable in domains that require high throughput and can tolerate specialized infrastructure. But in our eyes, the future of deployment depended on building smarter vehicles, not smarter pavement.

THE OTHER GUYS

Not every team in Demo '97 was tasked with building a full system like CMU or PATH. There were other participants — Toyota, Honda, Ohio State, Lockheed Martin, and Eaton — who each brought something to the table. They filled out the program, gave variety to the demos, and showed alternative directions the technology might take. But if I'm being honest, they were more like the undercard fights before the main event. The spotlight was always going to land on PATH and us.

Toyota brought what was called an "evolutionary scenario." Instead of diving headfirst into full autonomy, they focused on warning systems and driver-assist features: obstacle detection, lane departure, blind-spot alerts, and early versions of adaptive cruise control. Theirs was an incremental vision — safety enhancements and partial automation as stepping stones toward something bigger. It wasn't bad. In fact, it looked a lot like the systems that dominate the market today. But compared to the complexity of a full multi-vehicle, sensor-driven autonomy stack, it felt cautious.

Honda took a similar tack. Their "control transition scenario" was a two-car demo focused on shifting gracefully from manual driving into automated mode. The idea was to prove that handoffs between human and machine could be safe and reliable. That's still an issue people wrestle with today, so the foresight was good. But in San

Diego, surrounded by 15 ton transit buses driving at highway speeds and passenger vehicles executing tactical maneuvers, it came across more like a concept sketch than a headline act.

Ohio State University (OSU) was more intriguing. They experimented with radar-reflective pavement tape — a material that could serve as a cheaper, easier alternative to magnets embedded in the road. Only four miles of demo lanes were equipped with it, but it offered a glimpse of a hybrid path forward: minimal infrastructure upgrades combined with onboard intelligence. If you had reflective tape in place, the vehicles could run tighter headways or higher speeds. If you didn't, they'd fall back to their baseline autonomy. It wasn't a bad idea. In fact, of all the alternatives, it seemed the most plausible middle ground.

Lockheed Martin also participated, though on a much smaller scale. The irony was rich: the same company that had helped put a man on the moon couldn't muster the intellectual horsepower or resources to field a competitive self-driving platform. In truth, their demo was the work of essentially one guy trying to hold it all together on a shoestring. I knew him personally. He was a good guy, a smart guy, and I had even helped him in the past. But Lockheed didn't back him with what he needed, and it showed.

Finally, Eaton (Vorad) leaned into collision-warning technology. Their truck and target-car demo wasn't about full autonomy at all. It was about showing that radar could prevent rear-end accidents — an important, practical safety feature. It wasn't flashy, but it was the kind of thing you could see getting deployed in the near term, and in fact, pieces of that technology made their way into commercial trucking systems soon after.

All told, these demos were worthwhile, and it meant something to have Toyota, Honda, and GM all on the same stage. It showed that automated driving wasn't just an academic curiosity anymore — it had captured the attention of the world's biggest automakers. But the truth is, none of them matched the scope of what PATH and CMU put on the highway. We were showing full systems, with sensing, decision-making, and control stretched across multiple vehicle types. They were showing pieces. Good pieces, useful pieces — but pieces all the same.

DEMO '97 – THE MAIN EVENT

Demo '97 officially ran from August 7–10, 1997. The eyes of the U.S. Department of Transportation, Congress, and the press were on us. Secretary of Transportation Rodney Slater came to see it firsthand. Industry heavyweights, academics, and policymakers crowded the site, ready to judge whether automated driving was a pipe dream or a glimpse of the future.

To be clear, the runs weren't free-form. They were scripted demonstrations — but the vehicles still drove themselves. We told them when to attempt a maneuver, but the actual steering, throttle, braking, and obstacle avoidance were done autonomously. That was the only way to showcase the full range of capabilities in a safe, repeatable way.

Even with all the prep, there was always a sliver of fear. What if the GPS satellite constellation overhead wasn't favorable that day and we lost lock? What if a sudden downpour rolled in and blinded the sensors? What if — as I used to joke — a truck full of chickens overturned and flew into the lanes? You couldn't predict everything, and we knew it.

I remember that first morning vividly. We met in the giant Quonset hut for the pre-demo briefing, everyone wearing our sharp red, white, and blue team shirts donated by Houston Metro. Orders were given, roles reviewed. We even ran an early warm-up test to settle nerves. Everything worked. Then it was time for passengers. While the bigwigs in suits gave speeches about the promise of AHS, we sat quietly in our cars with the A/C blasting, waiting for our cue.

Even though I was the leader of the demo team, my role during the actual runs was a little unusual: I was one of the narrators on the transit buses. That meant I wasn't embedded with the participants during the run-up to the demo the way you might expect. I greeted them as they boarded, gave my spiel about autonomy, answered questions, and let them see the system at work. All the logistics — getting people to the site, giving them background about the technology, moving them between vehicles — were handled by others. My job, even on demo day, was still the same — and exactly what I wanted it to be: stay focused on autonomy.

Being on the bus proved to be an advantage. Standing up, able to look around and glance out the windows, I could monitor everything in real time. I could tell instantly when things were running smoothly and when something had drifted a little off-script. Our radio protocols worked exactly as they needed to at the start and at the end, and on the rare occasions when something wasn't behaving quite right, the systems we had built told us. And more importantly, we knew how to interpret what was happening and what the mitigation was.

A TYPICAL DEMO RUN

For all the politics, budgets, and big-picture arguments surrounding the Automated Highway System, most people only cared about one thing: what does it actually do? You could talk about architecture and policy all day, but nothing answered that question as convincingly as putting someone in the passenger seat, closing the doors, and letting them feel the system at work.

We'd rehearsed that "typical" run countless times. It became the backbone of the entire event — the sequence we relied on for senators, DOT officials, journalists, executives, and fellow researchers. Each run unfolded the same way: a calm briefing in the staging area, a few minutes of carefully orchestrated motion at highway speeds, and then the quiet exhale afterward as we rolled to a stop and watched our passengers absorb what they'd just seen.

The Briefing

Before anyone even saw the vehicles move, the experience began in the staging area. That was where we introduced ourselves, gathered each group of visitors, and set the tone for what they were about to witness. Chuck Thorpe usually led the way. He had the perfect blend of professor, tour guide, and ringmaster, and he made even the most intimidating technology feel approachable.

Chuck would greet each passenger individually, learning their names, affiliations, and what they hoped to see that day. Some came from government agencies. Some from industry. Some from overseas programs trying to understand whether the United States had actually cracked the code on automated driving.

Before we walked them around the vehicle, Chuck gave them context. The AHS Consortium wasn't a single vision marching toward one perfect answer; it was a collection of competing philosophies. Some

groups believed that dedicated automated lanes were the future —
clean, controlled, high-throughput corridors where everything talked
to everything else. Others, including CMU, believed automation
needed to function in ordinary traffic, surrounded by ordinary
human drivers, without requiring the nation to rebuild its entire
highway network. We toed the party line and explained that, in truth,
the country might need both. Dense cities could justify dedicated
infrastructure; rural areas, where automated vehicles would remain
sparse for years, needed solutions that blended in seamlessly. But we
also made our own position clear: automation had to work even
when the world around it wasn't automated.

From there, Chuck would walk visitors around the vehicle, pointing
out the hardware. He started with the tiny black-and-white camera
tucked behind the rear-view mirror. From that spot, it had a clear
view of the road and access to the wipers and defroster — a small
detail, but essential in real-world conditions. That camera fed an
ordinary PC sitting in the trunk, running the same vision code that
had steered Navlab 5 on No Hands Across America. We gave them a
quick rundown of how RALPH worked, explaining how it detected
every linear feature running parallel to the road — lane paint,
shoulder edges, tar lines — and used them to build a stable estimate
of where the car should go.

Foreign visitors were especially fascinated by it all. They'd crouch down to photograph the camera through the windshield, then circle to the back to take close-ups of the PC in the trunk — snapping pictures of wiring, brackets, connectors, nearly anything they could capture. Some even leaned in so close that Chuck had to gently remind them the demo hadn't actually started yet. Their enthusiasm became a running joke among the team: if it had a cable coming out of it, somebody was going to photograph it.

Next came the steering actuator: strong enough to hold a lane and execute quick maneuvers, but intentionally weaker than a human driver so that taking over was as simple as grabbing the wheel. If you fought it for more than a couple of seconds, it yielded entirely. Safety wasn't just a constraint — it was the philosophy that shaped every decision.

Then we gathered the group at the front bumper and explained the radar system — a mechanically scanned unit measuring range, bearing, and closing speed. We pointed out that its field of view was intentionally wide by design, wide enough to "see" vehicles in adjacent lanes, notice targets as we approached curves, and even pick up cross-traffic at intersections. Left on its own, that wide view would happily lock onto guardrails or signs. But once we explained how the system fused radar data with RALPH's understanding of the lane, the passengers could see how it all fit together. That fusion allowed the car to judge what was directly ahead, ignore what wasn't relevant, and maintain a steady speed while automatically slowing behind slower vehicles.

Next we moved the riders to the rear of the car and pointed out a small scanning laser rangefinder integrated into the bumper that swept a narrow arc behind us — part of our early experiments using different sensing modalities on real roads. And on the roof sat the GPS antenna. Most people knew GPS as a new and somewhat rough navigation tool, but ours used a differential system accurate to a few meters. That was good enough to check the reported positions of other AHS vehicles and validate what communications were telling us.

Once the outside tour was complete, we loaded everyone into their assigned vehicles. We made sure each passenger was buckled in, comfortable, and able to see the main display. On the buses, where there were no seatbelts, we reminded everyone to stay seated and hold on to the railing.

Inside the cabin, the visitors saw two displays. One was just for demonstration — it showed the front camera feed overlaid with what RALPH, the radar, and the laser sensors were detecting. The other resembled what a production interface might someday look like: system status, mode transitions, check-in results, and fault indicators. There was even a head-up display projected onto the windshield for the driver, just as future vehicles might have provided.

By the time Chuck finished the walkthrough, passengers understood exactly what they were riding in — not a science-fiction prototype, not a toy, but a real vehicle with real sensors running real code. It demystified the technology just enough to make what came next feel grounded rather than magical.

Then we closed the doors, radioed that we were ready, and joined the queue toward the highway entrance — the moment when the pre-demo briefing ended and the real demonstration began.

The Demonstration

What struck me most about the Automated Highway System demo wasn't the spectacle of it — the sedans, the buses, the minivans, the carefully choreographed runs — but how ordinary the whole thing felt once it was actually working. From inside the vehicle, automation didn't look like science fiction. It looked like a car quietly doing everything it was supposed to do.

The sequence began the moment we merged onto the dedicated AHS test lane. As we picked up speed, the system performed its own internal checks: lane visibility, radar lock, and overall vehicle status. Only after the vehicle confirmed all those boxes were green did it give the driver permission to hit the start button. And when he did — a simple press of the green "AUTO" switch — the wheel relaxed in his hands. The computer took over.

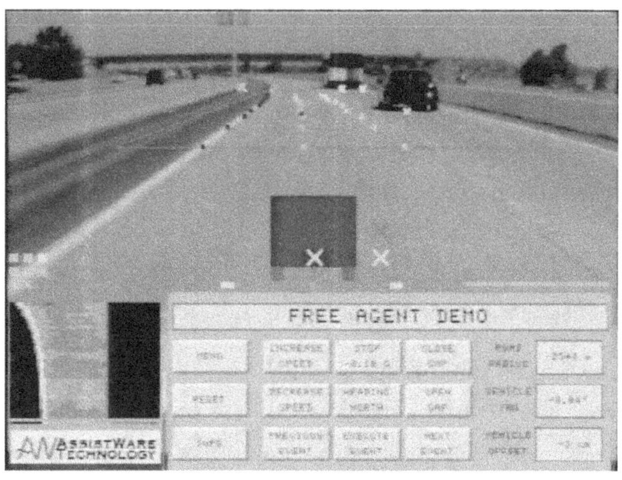

On the center display, the vision system painted its understanding of the world. It wasn't just looking for painted lane markers, because I-15's HOV lanes weren't exactly laid out with a ruler. Instead, RALPH searched for any linear features running parallel to the roadway — tar cracks, shoulder edges, the faint remnants of old paint — and stitched them together into a stable lane estimate. Blue dots marked the lane boundaries; a small crosshair floated ahead, showing the system's steering target. Radar data showing obstacles was layered on top, including a bright dot representing the back of the transit bus ahead of us.

We were nominally set to run 50 miles per hour, but automation isn't about the number you type in — it's about the vehicle in front of you.

That bus happened to be traveling at forty-five, so we simply fell in behind it and held the proper following distance. The bus behind us, though, had other plans. Not wanting to settle in at forty-five, it initiated an autonomous lane change and moved around us. All of the AHS vehicles — buses, sedans, minivans — ran essentially the same vision and control stack, with only modest differences in sensing hardware. Once the trailing bus completed its pass and slipped back into our lane, the lead bus accelerated. We did the same automatically, matching its speed without a single human input.

A moment later, the cars in the adjacent lane began their part of the choreography, accelerating and passing the bus convoy. During this segment, we deliberately maintained a fairly generous forty-meter gap — wide by design — because the whole purpose of the demo was to show how automated vehicles could blend safely with manually driven traffic. Loose spacing preserved stable vision estimates, allowed the system to react comfortably to speed changes, and demonstrated that even without tight coordination the system could nudge highway throughput above typical human-driven levels. A manual lane might carry two thousand vehicles per hour; our setup pushed that number toward twenty-five hundred. Nothing dramatic

— just a quiet proof that automation didn't have to break the world to improve it.

We also wanted to show courtesy. At one point, the Navlab 8 minivan surged up quickly behind the lead sedan. Sensing the rate of closure, the front car simply moved aside — an automated version of the thing most polite drivers do instinctively. Once cleared, the convoy picked up the pace again, the sedans now talking to each other via short-range communication. That allowed them to close their gap slightly, demonstrating how cooperative control could incrementally increase efficiency once enough automated vehicles shared the road.

The obstacle-avoidance segment came next. Far ahead, just barely visible to the human eye, we had placed a standard orange construction barrel in the lane. Even though the barrel wasn't metallic and gave a weak long-range return, the radar still detected it at about eighty meters — long before a human driver would have reacted. That was more than enough distance for the system to plan a smooth lane change or, if the adjacent lane had been occupied, to brake to a quick but controlled stop.

The car behind us, however, was following closely enough that it didn't have the luxury of waiting for us to move before reacting. So as soon as our radar tagged the barrel, our vehicle broadcast its location to any AHS-equipped vehicle in range. The trailing car compared that broadcast to its own GPS position, recognized that the obstacle lay directly in its path, and changed lanes without ever needing to see the barrel itself.

It was a small moment, but an important one — a glimpse of how cooperative sensing would eventually allow automated vehicles to share awareness, not just space.

And then, just as suddenly as it began, the demo drew to a close. The script called for a smooth, controlled stop at the end of the test track. The vehicles coordinated their deceleration, settled into a uniform glide, and rolled to rest as if the entire sequence had been one continuous breath.

From the outside, it may have looked like a choreographed performance. From the inside, it felt like the future behaving exactly the way the future should — quietly, confidently, and without any unnecessary drama.

Reactions and Reflections

Throughout the demo, I felt a calm confidence — not just in the technology, but in the people around me. The team supporting the live runs was steady, responsive, and completely prepared. When something needed attention, they handled it. When something went wrong, they already knew what to do. Putting my trust in the team, letting them do what they were trained for, gave me the freedom to focus on my part and to actually experience what we'd built.

It was the same feeling I'd had as a teenager playing sports, and would later feel as a coach: find the best people, train them well, trust them, and when game time comes, just let them play. San Diego was one long game day. The pressure was high, the stakes were real, and the audience was full of people whose opinions mattered. And yet, when it counted, the plan held — the systems did what they were supposed to do, and the people did too.

When the first official run finished cleanly, I felt a wave of relief wash through me. Just getting through one was enough validation. But the momentum only built from there. Across four days, roughly eight hundred people rode in our vehicles, logging more than 6,000 passenger-miles under automated control.

Their reactions were as varied as their backgrounds.

Most were amazed — genuinely wide-eyed. Many stepped out of the vehicles firing off questions faster than we could answer them: How does it know the lane? Could it handle rain? What if the lead car fails? How soon could this be commercialized? They'd gesture at the screens, the radar bumpers, the actuators, trying to mentally reconstruct how the whole thing held together at 55 miles per hour.

Others took the opposite view. They shrugged and said it felt… normal. Almost disappointingly normal. One visitor told me, "Honestly, it just seemed like a well-behaved car," as if that wasn't the entire point of the exercise.

Foreign visitors were their own category. Many of them would finish the ride and immediately break into their native languages, excitedly comparing notes with the partners they'd traveled with. I couldn't understand the specifics, but the tone was unmistakable: a mix of surprise, calculation, and sometimes disbelief. A few took more pictures after the ride than before, leaning into the cabin to shoot close-ups of the display, the buttons, even the stitching on the steering wheel, as if the details might reveal some hidden trick.

But no matter the reaction — awe, curiosity, or the quiet "what's the big deal?" — they all saw the same thing: vehicles merging, following, braking, and accelerating at highway speeds with headways as tight as 6.5 meters. They saw obstacles detected and avoided in real time, and obstacle locations shared using our early, almost primitive, vehicle-to-vehicle communication system.

For me, that stretch of I-15 became more than a test site. It was proof that the years of work — the long nights, the failed runs, the endless debugging — could be distilled into something an ordinary person could sit in, ride through, and walk away from saying, "Yeah. I can see this future."

VIP RIDERS

For the most part, the demos carried VIPs whose names and titles were lost on us. They were clearly important, but to us they were just faces in the seats. Rodney Slater, though, was different. We knew the Secretary of Transportation would ride, but not whose vehicle he'd be assigned to. It could've been us, PATH, or someone else. I was thrilled when I found out he'd be on one of the transit buses — my bus. Maybe the higher-ups thought our demo was the strongest, maybe they felt a transit bus fit DOT's multimodal mission, or maybe they just knew we'd whine and cry if we didn't get him (and they weren't wrong, lol). Whatever the reason, it was exciting to have him aboard.

Slater was engaged, attentive, and asked good questions. He got the full show: obstacle detection, lane-keeping, adaptive cruise, and coordinated maneuvers. And it all worked flawlessly. Watching him ride, knowing the Secretary of Transportation was seeing our work firsthand — and seeing it succeed — was one of the proudest moments of my career. In fact, it may have been too good, as I'll explain a little later.

Meanwhile, PATH's platoon of eight Buick LeSabres demonstrated their infrastructure-driven vision: a mechanical train of cars locked at tight spacings, gliding smoothly along. It was technically impressive, but also underscored the contrast.

RESULTS

By the end of the AHS Demo, both infrastructure-centric and vehicle-centric automation had proven feasible. But only one approach felt practical. What people remembered were the Free Agents — buses weaving through obstacles, minivans changing lanes, passengers gasping at how normal it all felt.

The stats backed it up. Out of 100 runs, 91 were successful. Measured in people-miles, we logged 6,378 out of 6,400 without incident, a 99.7% record. The buses were flawless: 40 runs, 40 completions. Most of the nine misses came down to minor hiccups — a GPS lock that lagged, a radar return that confused the system, or a safety driver disengaging out of caution. Only once did a run fail outright, when GPS never engaged at startup. Even then, the vehicle stayed under control. When engaged, the system did what it was supposed to do: lane-keeping, adaptive cruise, obstacle detection, and cooperative maneuvers at highway speeds, with real people onboard. The stats proved the point. But the bigger story was what they meant — for DOT, for industry, and for the future of how cars might drive themselves.

Within the consortium, the AHS Demo cemented the impression other teams already had of us: sharp, a little brash, and willing to get things done. On the technical side, there was respect — sometimes grudging, but real. For the DOT staff, we were the new group with a different style, less polished but highly effective. Our attitude might not have been their preference, but it didn't hurt us. If anything, it made clear there was more than one way to build credibility.

We weren't thinking in terms of reshaping policy or setting national direction. It was more instinctual: prove our vehicle-centric approach worked, and prove it worked in front of the world. But by standing up against the incumbent infrastructure-first model — the one many assumed was inevitable — we showed another path forward: faster, leaner, and scalable. Cars that could think for themselves, even in a limited way, shifted the debate about what autonomy might look like in practice.

Immediately after the final run, the mood shifted from tension to celebration. Everyone gathered in the Quonset hut as the program leaders offered a few closing words. The rivalry with PATH was over, and the atmosphere felt like a football locker room after a hard-fought season — grudges faded, respect remained. Everyone was elated; it felt like a triumph. But I think we were only beginning to grasp what we'd really accomplished across all the teams. Individually and collectively, we had created something that would echo for decades to come. We had created the future.

For me, the demo was validation. After months of non-stop work, juggling code, politics, buses, and babies, we had shown what was possible — and for the first time, the world at large had seen it. Barb felt that pride too, but at home that pace meant she carried more than her share. I was learning — sometimes too slowly — that leadership wasn't only about code, but about the people who were counting on me.

Barb's Perspective — On Pride

I've always been proud of Todd's accomplishments. Each new milestone made me happy. I used to joke "happy husband, happy life," but it was true — his joy in achieving things made me happy too.

All told, CMU's share of AHS probably cost around $2 million — a few years, a handful of vehicles, and a team you could count on one hand. Since then, the auto industry has poured over $20 billion into the same pursuit. GM and Tesla each have spent over $10 billion, and Ford and Volkswagen sank billions into Argo. That's more than 10,000 times what we had at CMU. Yet the core challenge remains the same: making vehicles think for themselves, safely and at scale.

The AHS demo was never a scored competition with judges, trophies, or an official "winner." It was a federally backed showcase — proof of feasibility for DOT, Caltrans, and the public. The legislation set the stage, the consortium delivered, and history did the sorting. And history's verdict was clear: our team proved what billions are still chasing.

MEDIA AND PUBLIC RECEPTION

If the AHS demo was designed to show the world what was possible, the world showed up. The media presence was staggering. More than 120 outlets were represented, from the big three networks — ABC, CBS, and NBC — to CNN, MSNBC, and PBS. NPR's *All Things Considered* ran segments, and print reporters from *The New York Times, Washington Post, Wall Street Journal, USA Today,* and *Newsweek* filed stories. The Associated Press carried the event nationwide, and the reach went even further: international coverage spanned at least 12 countries across five continents.

Watching the reel of TV clips today is a time capsule of the era — anchors marveling at "robot cars," footage of buses gliding down I-15, sound bites from government officials promising a safer future. Even David Letterman joked about the demo during his monologue on his show. To us, it was surreal. We had spent 18 months buried in code, debugging in Quonset huts and sweating in 100-degree heat. Suddenly, those same machines were on the evening news, broadcast to millions. For a tiny team of researchers, it was both validation and a little bit of whiplash: one week we were soldering wires and cursing GPS locks, the next we were in *Newsweek*.

The public response was equally striking. Across all the demos, 2,850 people rode in automated vehicles during those four days. Many stepped off wide-eyed, remarking how "normal" it all felt. That may have been the biggest achievement of all: not just proving the tech worked, but making autonomy feel real, safe, and inevitable.

One of the coolest touches we added — and something we never really cleared with the AHS bureaucracy — was creating trading cards for the vehicles. Yes, actual baseball-style trading cards... for robots. Each Navlab, from 1 through 10, had its own glossy card with a picture on the front and stats on the back: make, model, platform quirks, and the kind of specs that only a robotics nerd could love. After every ride, passengers got to take one home.

It was half outreach, half inside joke. Who else can say they own a rookie card of a self-driving bus? Or flip over a minivan's "stats" like it was batting .320 in the National League? For all the seriousness of the demo — the VIPs, the press, the politics — those cards gave it a little wink. They turned cutting-edge robotics into something you could literally stick in your wallet. And judging from the smiles, I think people loved the fact that, for once, science handed out souvenirs cooler than a brochure.

ACADEMIC RECOGNITION AND IMPACT

The recognition that followed also carried weight. Our team received the Allen Newell Award for Research Excellence, given annually by CMU's Computer Science Department to projects that exemplify Newell's style of research. His philosophy was simple but demanding: good science responds to real problems, good science is in the details, and good science makes a difference. The AHS work checked all three boxes. We weren't chasing theory for its own sake — we were solving real problems of driving on real roads. Every result was grounded in painstaking details, from demo scripting and safety protocols to ACC algorithms. And most of all, it made a difference: demonstrating to the entire world that self-driving vehicles of all types could operate safely and reliably at highway speeds.

I knew Allen Newell was a giant in computer science, but I had no idea who else had won the award until I looked back later. The roster was humbling: Raj Reddy's first speech recognition systems, early deep learning that conquered chess, Omead's helicopter autonomy work, surgical navigation robots that Takeo helped develop. What makes the Newell Award so uniquely "CMU" is that it doesn't just celebrate theory or prototypes — it celebrates both. Hard science and real systems and I'm very proud we were counted among that lineage. It was clear that in both the eyes of the public and the academy, the AHS demo had crossed a threshold: autonomy was no longer just research; it was real.

And the legacy didn't stop with awards. CMU was a crucible. Even people who weren't directly involved in our demo were exposed to the technology, the ideas, and the culture. Many went on to staff the first generation of teams seriously pursuing autonomy — from DARPA Challenge efforts to early autonomous-vehicle startups. The DNA of that demo spread far wider than the handful of buses and minivans that ran down I-15. And truthfully, by that point Dean and I could have gone almost anywhere and continued our research.

Even now, when I drive my Tesla, I sometimes smile. On highways, its adaptive cruise doesn't feel much better than what I built in 1997. Sure, it handles a wider variety of conditions, but in the pure driving experience — smooth following, natural braking, human-like acceleration — my system would still hold its own. At the time, it felt like we were just scrambling to meet deadlines. With perspective, I can see that Dean and I were laying down an architecture for autonomy that would last for decades.

And that's really the legacy. Not just a week in San Diego, but a blueprint: smarter vehicles, not smarter pavement; incremental deployment, not trillion-dollar construction projects; safety nets built in and tested under pressure. It was a philosophy that started with the Navlabs, proved itself on I-15, and continues to shape the autonomous systems rolling out today.

Looking back now, it's clear that what we were building wasn't a finished system — it was a beginning. From the early days of the DARPA Autonomous Land Vehicle (ALV) program through the Automated Highway System nearly a decade later, sustained government funding made it possible to explore ideas industry wasn't yet ready to touch. Graduate students and researchers were given the freedom to build real systems, take real risks, and test fragile technology in the real world — driven more by optimism than certainty.

In that sense, the work mattered less for what it immediately delivered than for what it enabled. Those publicly funded programs created space for learning by doing — for ideas to mature outside the constraints of near-term commercial viability. Over time, as the technology hardened and the path forward became clearer, industry stepped in and carried it the rest of the way. Like the early Internet, the arc ran from government investment, to academic experimentation, to private adoption at scale.

The true legacy of ALV and AHS wasn't a single vehicle or demonstration. It was the people shaped by the work, the confidence gained

from trying, and the belief that the future was worth pursuing —
even when success was far from guaranteed.

AFTER THE DEMO

After the adrenaline rush of San Diego, we finally took a breath. For
the first time in eighteen months, there wasn't another impossible
deadline looming. The demo had worked. The pressure lifted. And
for a little while, we allowed ourselves the rare luxury of slowing
down.

When the AHS equipment was packed up and shipped back to Pittsburgh, life quieted considerably. After months of nonstop intensity, it
felt almost disorienting to have no crisis to solve. For me, it was a
chance to be home more, to spend time with Emma, and to catch my
breath as a new dad.

There was one short trip down to Houston, where we staged a
smaller demo with Houston Metro. This one focused only on transit
buses and a single vehicle, and I spent about a week on-site. Beyond
that, the calendar was clear until Christmas.

After the new year, though, we landed funding for a project I thought
was genuinely innovative. Rear-end collisions with transit buses were
— and still are — a serious problem. Our idea was straightforward:
mount a system on the back of a bus that combined radar, strobe
lights, a variable message sign, and an ear-splitting horn. If the radar
sensed a vehicle approaching too quickly, the strobes would flash, the
sign would warn "Slow Down," and the horn would blast. It never
went beyond the prototype stage, but I thought it could have made a
real difference.

Finally, recall that the Navlab 5 never made it into the AHS spotlight.
Once the new vehicles and buses came online, it sat idle, the NHAA
road trip already behind it. When the AHS program wound down,
Delco reclaimed it and scrapped it. We had drilled holes, cut panels,
and carved it up so thoroughly that it couldn't be resold or reused. On

paper, it was gone. But in truth, Navlab 5 had already crossed into history.

A decade later, in 2006, it was inducted into the Robot Hall of Fame for the No Hands Across America trip. For all its quirks, it had done something unforgettable — and history remembered. Like the Wright Flyer or the first iPhone, its value wasn't in the hardware left behind, but in the leap it represented. Even a vehicle scrapped for parts had earned its place as a symbol of what was possible.

WHAT CAME NEXT

By 1998, the bigger picture had shifted. The Automated Highway System program was being wound down. Part of it was politics, part of it was funding, and part of it was the natural cycle of government research programs. San Diego had proven the point: autonomy wasn't science fiction anymore. It was real. And once that was clear, the Department of Transportation seemed to decide it was time for commercial organizations to take the next step.

I've always liked to think our demo had something to do with that. We showed how far the technology could be pushed under real-world conditions. Maybe the government looked at it and thought: "We don't need more research. We need industry to make this real." In hindsight, they were probably right. The seeds were planted in the 1990s, but it would take another decade before the first driver-assist systems began appearing in production cars.

Sure enough, the path forward unfolded almost exactly that way. The Free Agent philosophy lived on — first in adaptive cruise control and lane-keeping, then in the DARPA Grand Challenges of the 2000s, and eventually in the rise of dedicated autonomous-vehicle companies. The infrastructure-heavy approach never escaped the realm of research papers and polished prototypes, but the CMU vision became the practical baseline.

Now, nearly thirty years later, you can finally look back and call the demo for what it was — and if you're keeping score, CMU won. Not by having the biggest budget or the tightest control, but by choosing the approach that actually endured: put the sensors on the vehicle, let the vehicle sense the world for itself, and make it figure things out in real time. That was the philosophy baked into ALVINN, RALPH, and eventually NHAA and AHS — the idea that autonomy had to live on the car, not on the roadside.

To be fair, the race isn't over. The field is still evolving, new architectures come and go, and no one has fully solved autonomy everywhere, all the time. But at least for now, the trajectory is undeniable. The lineage that runs from our duct-taped Pontiac minivan — perception on the vehicle, real roads, real uncertainty, real adaptation — is the one that carried forward. What we proved in 1995 and then again in 1997 has been rebuilt, industrialized, and deployed on tens of millions of vehicles across the globe.

So if you're scoring the demo with the benefit of hindsight, it looks a lot like we won.

For me personally, AHS winding down meant something else too: I could feel my own time at CMU coming to an end. The program that had consumed me for nearly two years was over, and with it the momentum that had tethered me to academia. The Ph.D. had been the capstone; AHS was the encore. Together, they gave me everything I needed: technical chops, a trial by fire in leadership, and the confidence that our work could stand on the national stage.

Just as important, I felt content. I didn't have a burning desire to keep chasing more papers, more projects, more research for its own sake. I had set out to make a mark in academia and I believed I had done that. It was good enough. It mattered. And with that, I was ready to walk away and see what might come next.

During the fall of 1998, Dean and I were starting to think seriously about leaving CMU and building something of our own — how to take the momentum of the AHS work and translate it into something lasting through AssistWare. One of the unexpected dividends of the demo was that it had already planted the seeds for our first real customers. Delco Electronics, a key player in the consortium, bought four of AssistWare's lane-tracking systems. GM went a step further, hiring us to help integrate the technology into their human factors testing platforms.

It was surreal. Just months earlier, we had been hustling inside CMU to keep the AHS program on track. Now the very companies that had been our partners in the consortium were turning to us directly for solutions. Another boost came from CMU itself, which agreed to subcontract work to AssistWare so Dean could finish out a research project he had been leading. That support gave our young company a crucial bridge of stability and credibility at exactly the right time.

And those early contracts did more than validate our work. They proved there was a market for it, and they gave AssistWare the first real hint of legitimacy. But the thought of leaving brought up the same sort of feelings I'd had when I first came to Carnegie Mellon. I was excited and scared for what came next, but I knew I had to do it. We had a new baby. I had a window of opportunity. And the reality was simple: I was about to walk away from something relatively secure and step into the unknown again — except this time the stakes weren't just mine. They were ours.

The next chapter wasn't going to be in a lab or a classroom. It was going to be out in the world — building things, starting companies, and finding out whether I could take what I'd learned on the road, literally and figuratively, and turn it into something lasting. But that's for the next book.

REFLECTIONS

The AHS prep consumed my life. Long days blurred into long months. Our daughter Emma was born in May and Barb carried far more than her share at home. When we all went to California together, it was the first time we tried to blend the chaos of a newborn with the chaos of a national research project.

We stayed in a condo on the beach, stealing moments of normalcy between the stress. One ritual was grabbing food from Rubio's, the local fish taco spot. In the middle of everything — Emma's arrival, twelve-hour days, endless debugging — those tacos were one of the few small pleasures that cut through the exhaustion. To this day, I measure every fish taco I try against Rubio's. Thirty years later, I finally went back and rated them an 8.8 out of 10 — the highest score I've ever given. That tells you just how good they were, and how much those tacos became a touchstone for me.

But exhaustion also made us overreact. One night Emma had some small issue and Barb and I, overtired and overprotective, rushed her to the ER. It turned out to be nothing at all — just two new parents stretched thin, trying desperately to do everything right.

Not every memory from that summer was stressful, though. One afternoon, I looked out our condo window and saw that Chuck's wife, Leslie, and daughter, Hannah, had written "Emma" in the sand. It was a reminder that even in the middle of a national demo, people around us were rooting for our family too.

But the real hero that summer was Barb. She was only six weeks post-partum, thousands of miles from home, with no family support nearby. Yes, Chuck's wife Leslie and Dean's wife Terry stepped in and helped when needed, but it had to be hard. Barb spent her days with a brand-new baby, often without a car, and little to do beyond walking the beach. On the surface, it might sound idyllic, but anyone who has raised a newborn knows it isn't easy. And yet she never complained — not once. She simply embraced the role of mother and partner, living in the moment and supporting me without hesitation. She was the rock of our little family.

Barb's Perspective — On Todd's Long Stretches of Work

Todd being buried in AHS or NHAA never really bothered me. I had my own full schedule at Pitt and with my internship, and later with Emma. I spent free time doing activities with her, or with Pitt and CMU friends. The only time I really felt lonely was when he was in Denver for his Fellowship. Our little house was hot, the days long, and the evenings could feel very quiet. But otherwise, I stayed busy enough that his work didn't weigh heavily on me.

On the professional front, the demo also taught me lessons I didn't fully appreciate until later. At the time, I was bullheaded — convinced our way was the only right way, frustrated by bureaucracy, unwilling to see the value in compromise. If someone disagreed with me, I pushed harder. Only years later, running my own companies, did I realize that sometimes bureaucracy can't be eliminated; it has to be navigated. The real skill was in seeing the other person's perspective and finding ways to create win-win situations. That shift — from "me-first" stubbornness to at least trying to understand the other side — became one of the things that helped me as an entrepreneur, and maybe even more as a husband and father.

At first, that lesson carried over into my companies, but it also shaped me at home. I won't claim I stopped being a self-centered only child — I'm still wired that way in plenty of respects — but I became more conscious of it. I learned to pause, to listen, and to realize that sometimes the other side of the table wasn't the enemy; it was the partner you needed to make progress. And that realization, forged in the heat of the Automated Highway System, carried me further than I could have imagined.

LOOKING BACK, MOVING FORWARD
WHERE THE FUTURE TOOK HOLD

"What lies behind us and what lies before us are tiny matters compared to what lies within us."

— *RALPH WALDO EMERSON*

When I look back on my years at Carnegie Mellon, I see more than a degree, more than the experiments, and more than the projects that made headlines. What I really see is a period of transformation — professionally and personally — that shaped both the person I was and the one I was becoming.

I had arrived in Pittsburgh as a kid from Indiana, raised on hard work, accountability, and the belief that persistence mattered more than polish. At CMU, those values were tested and stretched. The problems were bigger, the stakes were higher, and the competition was tougher than anything I had ever faced. CMU didn't just sharpen my skills — it taught me how to build momentum. One problem solved, then the next. Not glamorous but steady. That rhythm shaped everything that came after.

Professionally, CMU was both a proving ground and a launchpad. NHAA taught me boldness — the courage to try something no one had done before. My thesis taught me discipline — the patience to push through the details even after the excitement fades. AHS taught me leadership — the responsibility of guiding a team toward a goal bigger than any one person.

And somewhere in the middle of all that work, I began to understand something about myself: building things wasn't just ambition for me — it was how I showed I cared. That drive to make something real, to make it work, was as much about the people around me as it was about the technology itself.

Together, those projects showed me that breakthroughs don't come from a single lightning strike. They come from daring matched with endurance, from taking an idea that might work and grinding until it does. I didn't feel like I fully belonged when I first arrived at CMU, but by the time I left, I knew I'd become exactly who I was supposed to be.

More than anything else, my academic career simply felt complete. I never once thought about going back for more. The Ph.D. was the goal, the capstone, and I had no desire to chase another one. When I first came to CMU, it was simply for the degree — I had no idea that No Hands Across America or the Automated Highway System even existed. Those opportunities unfolded along the way, and they changed the trajectory of my life. But the Ph.D. was the spark that lit it all. From that point, my classroom would be the world.

Personally, CMU tested more than my code — it tested my balance. The lab pulled long hours, but at home Barb carried equal weight, steady and patient, even when the money was tight and the outcomes uncertain.

I'll never forget standing in the hospital holding Emma for the first time, feeling both awe and responsibility crash over me in the same breath. That moment made everything more real: this wasn't just about research anymore. Barb's faith and that little girl's arrival meant the stakes were deeper than any advisor or committee could impose. The experiments might have been mine, but the journey was ours.

Barb's Perspective — On Our Lives During This Period

Those years feel like a blur of adventure. Life was simple: we didn't have much, but we made that little row house into a home. It was exciting, it was new, and it went fast.

In hindsight, I don't think we missed much. We were stepping into the season of raising kids, and I was happy at home doing the 'kid stuff.' Todd was around enough to be part of that. The balance felt right for where we were in life.

Our biggest accomplishment as a team was moving to Pittsburgh and thriving. We could have quit and left when things got hard, but we didn't. We stayed, made it work, and built something together.

MEANWHILE, IN THE WORLD

While I had my head down in academia, immersed in research, the world outside kept moving. One of the first moments that really broke through for me came in 1990, when the U.S. launched its first strikes against Iraq after the invasion of Kuwait.

I can still picture Barb and me sitting on our brown couch, watching it unfold live on TV, trying to process the surreal reality of war beginning halfway around the world. I wasn't thinking about geopolitics — I was fascinated by technology like the stealth fighters we were suddenly seeing on the news. But now I see it was more than just technology. Back then, I saw it through the lens of technology — stealth aircraft on TV — not the geopolitical shift it represented.

A few years later, in 1993, the first World Trade Center bombing shook New York City. I never imagined it as a prelude to 9/11, or that it would spark the kinds of global divisions and conflicts we now live with every day. It simply felt like an isolated incident — not the precursor we now know it was.

Not long after that came another surreal national moment: the O.J. Simpson chase and trial. I vividly remember standing in the common area of Smith Hall, where the Robotics Institute had moved in the mid-90s, watching the drama unfold with other students and researchers. Seeing a sports icon in a white Bronco crawling down the L.A. freeways, trailed by what seemed like every police car and helicopter in California, was as bizarre as it was unforgettable.

By the late 1990s, the mood had shifted. The Internet boom was exploding — Netscape, AOL, dot-com startups everywhere — and it was impossible to ignore. Dean and I knew that world wasn't our lane, but the energy around small teams building real companies absolutely rubbed off. It proved that a handful of people could create something meaningful. Academia was comfortable, and CMU offered prestige and freedom, but a startup offered a different kind of challenge — the chance to build something new and, maybe, to take care of our families in a way academia never could. And seeing what other people our age were doing, seeing what the Internet was inspiring, made me eager to thrust myself into the unknown of starting a company. It was the same mix of fear and conviction I'd felt before — scared of the leap, but knowing it was the right thing to do.

What stands out to me, all these years later, is how much of what seemed like "just another headline" was really the beginning of something much larger. The Gulf War, the World Trade Center bombing, the O.J. trial, the Internet boom — they were all early chapters in stories that would define the decades ahead. I barely noticed the big picture. But from today's vantage point, it's clear those moments weren't isolated blips. They were the opening scenes of what was to come.

THE LIMINAL SPACE

Looking back, CMU was the bridge between where I came from and the life that followed. It expanded my capacity to take on hard things — things I couldn't have even pictured growing up in Indiana. More than that, it was where I started the real work of becoming an adult. I arrived in Pittsburgh largely untested, and over those years I grew into roles I had only vaguely understood: husband, father, researcher, leader. Each step demanded more than I thought I had, and each one showed me I could rise to it.

The Ph.D. was never just about letters after my name — it was about building a foundation strong enough to carry us forward. It gave me the confidence to close one chapter and the courage to open the next. I walked in unsure if I belonged, and I walked out ready to shoulder the risks and responsibilities of the life Barb and I were starting together.

But the biggest question was still ahead: could what thrived in the lab survive in the real world? Leaving behind the safety net of advisors — and even the relative security of a prestigious academic path — meant stepping into a world with no guarantees. For the first time, the experiments weren't just mine. The risks, and the stakes, would be, too.

Barb's Perspective — On Leaving CMU

When Todd said he wanted to leave CMU and start his own company, I'll admit I was a little nervous. I didn't know much about startups, but I knew they weren't as stable as the job he already had. At the same time, I could see he was ready for something new. He'd gotten everything he could from CMU, and staying would've felt like standing still. Watching him through that time, I could see how much he'd grown — and in many ways, we began to grow up together. Even with our family just starting, I trusted his drive, and I knew we'd figure it out together.

I had gotten everything I wanted from CMU — skills, credibility, a Ph.D., and experiences that stretched me in every direction. But I didn't want a life of grants, committees, and tenure ladders. I was ready for something new. Entrepreneurship wasn't just about money, though that mattered. It was about testing whether the ideas we had built in the lab could stand on their own in the real world. Building a technology company wasn't only about code or algorithms anymore. It meant learning how to lead, how to see problems from a customer's perspective, and how to rally a team behind a vision when the path forward wasn't clear. It meant taking responsibility not just for ideas, for people, livelihoods, and outcomes.

The leap from student to builder — and from newlyweds to a growing family — was no longer theoretical. It was right in front of us. CMU had given me everything it could: the spark, the discipline, the confidence, and the proof that bold ideas could survive on real pavement. But what came next wouldn't be an experiment, or a demo, or a thesis. It would be the real world — customers, risks, responsibility, and outcomes that mattered far beyond a conference room or a test track.

Leaving CMU didn't feel like an ending. It felt like a beginning — the moment when all the momentum we'd built finally pointed forward. I didn't know what the next chapter would require — only that I was finally ready for it.

Book 3 begins there: the moment I stepped away from the safety of the lab and into the uncertain world of building something of my own.

THE NAVLAB FAMILY
OF VEHICLES

From 1986 onward, CMU built a fleet of experimental vehicles that came to be known as the Navlab family. Each one had its own quirks, sponsors, and mission. Some were rough prototypes held together by graduate-student ingenuity; others were polished enough to appear in high-profile government demonstrations. Together, they charted the arc of how self-driving research grew from a single hacked-up van to a diverse lineup of cars, minivans, and buses.

Navlab 1 (1986) — Chevy Panel Van: The original. A box truck converted into a rolling lab with racks of Sun workstations in the back and video cameras mounted above the cab. Slow and ungainly, but it proved you could strap computers to a vehicle and make it follow the road.

Navlab 2 (1991) — Army Humvee (HMMWV): Big, heavy, and rugged, this Humvee was designed to show that autonomy could scale to both off-road terrain and work at highway speeds. Outfitted with cameras, radar, and more compute power, it became CMU's first serious high-speed testbed.

Navlab 3 (1994) — Honda Accord (Dean Pomerleau's personal car): The budget version. A steering motor on the wheel, a camera suction-cupped to the mirror, and a laptop running off the cigarette lighter. Despite the garage feel, it proved autonomy could work in an ordinary passenger car.

Navlab 4 (1994) — Army Humvee (second HMMWV): A second Humvee, used for extended military research. With cameras and radar mounted up top and ruggedized electronics inside, it broadened CMU's ability to tackle off-road and unstructured environments.

Navlab 5 (1995) — Pontiac Trans Sport Minivan: Donated by Delco Electronics, this silver van became a legend. It carried the No Hands Across America trip from Washington, D.C. to San Diego, logging about 98.5% of the miles autonomously.

Navlab 6 (1996) — Pontiac Bonneville Sedan (My commuter vehicle): One of two identical Bonnevilles built for the U.S. DOT's Automated Highway Systems (AHS) program. Outfitted with video cameras, radar, and GM's by-wire throttle and steering interfaces, it was a showcase vehicle for automated driving on I-15 in San Diego.

Navlab 7 (1996) — Pontiac Bonneville Sedan: A twin to Navlab 6, with the same AHS package. Having two Bonnevilles made it possible to run coordinated demonstrations of automated lane changes, merges, and safe following at highway speeds.

Navlab 8 (1996) — Oldsmobile Silhouette Van: A larger platform used in the AHS program. It demonstrated extended automation tasks such as cooperative merges and high-speed following, with room for researchers and observers on board.

Navlab 9 (1997) — Houston Transit Bus: A full-sized transit bus adapted for the AHS demo. Cameras mounted on the front, radar for adaptive cruise control, and heavy-duty steering/brake actuators gave it full automation capabilities. It proved the technology could scale to mass transit.

Navlab 10 (1997) — Houston Transit Bus: A companion to Navlab 9, equipped with the same suite of cameras, radar, and computing racks. Together, the buses gave CMU the ability to demonstrate close headway driving at highway speeds with passenger loads, highlighting safety and scalability.

By the time of the **1997 San Diego AHS Demo**, Navlabs 6 through 10 were running alongside other project vehicles in front of government officials, media, and the public. The lineup illustrated not just the technical progress, but also the range of ambition: from a humble box van to buses and highway-ready sedans. The Navlabs weren't just vehicles; they were milestones, each one carrying a different piece of the puzzle that made modern autonomous driving possible.

ABOUT THE AUTHOR

Todd Jochem, Ph.D., is an engineer, entrepreneur, coach, and lifelong builder whose work helped launch the modern era of autonomous robotics. Raised in southern Indiana, he grew up straddling two worlds—basketball and technology—learning discipline in small-town gyms and curiosity in garages filled with electronics, rockets, and early computers. Those early passions eventually took him to the Carnegie Mellon University Robotics Institute, where he earned his doctorate and contributed to foundational research in perception and self-driving vehicles, including the landmark No Hands Across America project.

After leaving academia, Todd co-founded multiple robotics and computer-vision companies, helping bring advanced autonomy out of the lab and into industry. His ventures were acquired by major defense and technology organizations, and his work continues to influence systems that trace their lineage back to those pioneering days.

Todd is also a successful quarterback coach. He spent years on the football field—first at the youth level, then at the high-school level—bringing the same principles of preparation, accountability, and trust that shaped his own life. He helped guide teams to multiple state championships, contributing to a culture built on discipline, development, and leadership. Coaching became another arena where Todd found purpose: building teams, mentoring young athletes, and helping them discover how good they could truly become.

Through it all, family has been the foundation. Todd's professional life and coaching career grew in parallel with the life he built with his wife, Barb, and their children. The breakthroughs in the lab, the risks of entrepreneurship, and the victories on the field evolved alongside the quiet, steady work of raising a family and becoming the kind of husband and father he aspired to be. Writing this memoir series has given him the chance to see those threads together—and to honor the truth that neither his career nor his story makes sense without the people at the heart of it.

His series, *Dare to Be Great: A Life in Five Acts*, weaves these themes—family, grit, innovation, leadership, and the pursuit of excellence—into the larger arc of a life shaped by purpose and possibility.

Todd lives with his wife, Barb, and together they've raised a family that remains the center of everything he does.